Waste and Urban Regeneration

Waste and Urban Regeneration examines the Nanjido region of Seoul and its transformation from Nanjido Landfill to the World Cup Park, and its relation to the urban ecology within the context of the city's urban development during the late twentieth and early twenty-first centuries.

The study analyses the urban ecological meanings of the site's two distinct forms by consolidating them with the Lefebvrian urban theory and relational ecological theories. This book looks at environmental transformations and their link to South Korea's political and economic changes; how Seoul City controlled waste populations, the borderline characterisations of the inhabited landfill and its community, the regeneration of the landfill into the post-landfill park and site-specific artworks which explored the conflict between the invisible presence of the landfill's garbage and its history.

As one of the first accounts of a landfill and landfill-turned-park of South Korea, this study is a must-read for academics and researchers interested in waste management, ecology, urban studies, landscape theory and history.

Jeong Hye Kim is visiting professor of Seoul National University of Science and Technology with a primary research focus on architectural design and art in urban settings. Her subjects of research interest are the political and socio-economic relationship with the urban environment, post-traumatic historical spaces, sense of place[less]ness and ecological equilibrium. Translations include Hal Foster's *The Art-Architecture Complex* and Charles Jencks and Nathan Silver's *Adhocism*.

Routledge Research in Landscape and Environmental Design

Routledge Research in Landscape and Environmental Design is a series of academic monographs for scholars working in these disciplines and the overlaps between them. Building on Routledge's history of academic rigour and cutting-edge research, the series contributes to the rapidly expanding literature in all areas of landscape and environmental design.

For more information about this series, please visit: https://www.routledge.com/Routledge-Research-in-Landscape-and-Environmental-Design/book-series/RRLAND

Waste and Urban Regeneration

An Urban Ecology of Seoul's Nanjido Post-Landfill Park

Jeong Hye Kim

Routledge
Taylor & Francis Group

LONDON AND NEW YORK

First published 2021
by Routledge
2 Park Square, Milton Park, Abingdon, Oxon OX14 4RN

and by Routledge
52 Vanderbilt Avenue, New York, NY 10017

Routledge is an imprint of the Taylor & Francis Group, an informa business

© 2021 Jeong Hye Kim

British Library Cataloguing-in-Publication Data
A catalogue record for this book is available from the British Library

Library of Congress Cataloging-in-Publication Data
A catalog record has been requested for this book

ISBN: 9780367356408 (hbk)
ISBN: 9780429340871 (ebk)

Typeset in Sabon
by Deanta Global Publishing Services, Chennai, India

Contents

2 **Sanitary management in post-war Seoul**   59

3 **Nanjido Landfill as human habitat**   94

Figures

Tables

Preface

Cities have waste and waste treatment systems, without which they would not function properly and would end up becoming landfills themselves. Modern cities have more rigidly maintained physical and social cleanliness. As a consequence, the sanitation of public and private spaces has made places of waste equivalent to something dirty, immoral, wrong and thus inappropriate to the urban space. What's more, cities have even marginalised social groups that are regarded as politically and socio-economically inappropriate in order to maintain the cleanliness of social affairs. This extension of the material practice of sanitation into the social process is related to the shift in the concept of the *city* as a military machine that protects the inside from the outside world. In the new structure, human beings become a part of the *city* as *the urban* or a place where differences encounter, acknowledge and explore one another.

At the turn of the twenty-first century, both cities in the West and in the East began to regenerate old parts of the city. The logic behind this urban regeneration coincides with the cleansing and sanitisation of the modern city, and by extension, the aestheticisation of the city. In this regard, this study views urban regeneration in the new millennium as another form of waste treatment, and it connects the concept of waste management to the city's propensity to abandon anything considered inappropriate to the modern norms of the city, both materially and socially. In South Korean society, industrialisation, modernisation and urbanisation advanced one after the other in the condensed period of the 1960s and 1970s. Throughout this time, material and social sanitisation reached its peak. In this study, I use the Nanjido region's transformation from a wetland to a landfill (1978–1992/2001) to a park (2002–present) as a metaphor for the city's urban regeneration, showing how Seoul City carried out environmental sanitisation through social cleansing processes. Since this phenomenon of dual material and social layers has lasted throughout the history of modernisation, I examine the physical and social sanitisation policies of South Korea, particularly during the 1960s and 1970s. I delve into Seoul City's regulation of ragpickers—the symbolic figure of waste treatment—and control of the citizens' health and hygienic environment through fumigation with DDT, a chemical that symbolises the era's sanitisation.

We must look at the urban regeneration of the Nanjido region in the context of its regional redevelopment. On the one hand, Nanjido Landfill's transformation can be regarded as the transformation of brownfield (abandoned site of the industrial era) into greenfield (environmentally revitalised site of the neoliberal era), which is one of the most significant forms of landscape regeneration in the history of Seoul. On the other hand, the building of a digital technology-based smart city is a global phenomenon, particularly in the relatively lucrative developing cities in Asia like Seoul, Shanghai and Singapore. In the case of Sangam-dong's regional redevelopment, these two regenerations took place simultaneously; while the City planned to develop the old Sangam-dong town into the Digital Media City (DMC) in the north of the area, the government decided to transform the landfill into a park in the south. As such, both developments are intertwined and mirror each other, suggesting that the mechanisms of urban regeneration are an extension of the city's cleansing.

When I first commenced fieldwork, Nanjido Landfill was an invisible past; people no longer remembered the landfill or its transformation processes, and only the dichotomous concept (landfill-dirty vs. park-clean) lingered in their collective memories. Combing through references of the landfill felt like the process of uncovering the invisible past, or an archaeological excavation. Amongst the diverse methods I used to discover the buried past, e.g. interviews and document research, I also borrowed artists' aesthetic representations, which linked the material landscape and urban space to the psychological aspect of the place: the sense of place or placelessness.

As a multidisciplinary research based on history and sociology, this study traverses the disciplines of landscape, urban studies and ecological studies. Although the landfill's closure and transformations into a park are within the realms of landscape and geography, it is inextricably intertwined with the regional urban regeneration (coinciding with an international sporting event in particular) and the eviction and relocation of the landfill habitants, so this study extends its disciplinary boundaries into urban studies and ecological studies, while embracing aesthetic methodologies. By experimenting with this cross-disciplinary approach, I attempt to provide multiple lenses that will help us view our environment synchronically and diachronically.

I am most grateful to my supervisor Professor Peg Rawes for her inspiring encouragement, which led me to take on a poetic method in urban ecological studies, a new approach to this multidisciplinary research. I am indebted to my secondary supervisor Professor Murray Fraser for his sharp, timely and considerate comments, especially on the meanings of brownfield development in association with the building of smart cities. Professor Ben Campkin and Professor Harriet Hawkins' accurate review and comments were an immense help in articulating the structure and argument of the study. I am very thankful to Stylianos Giamarelos for peer reviews and immensely thoughtful comments and questions during the last stage of my research. I'd like to express my gratitude to Grace Harrison and Julia Pollacco of Routledge for their consistent support throughout the publishing process.

Without the 19 interviewees, this fieldwork-based research would have been impossible. I am deeply thankful for their precious time and endeavours to recall memories of their past in Nanjido, both good and bad. It is a great privilege to have friends and colleagues in the UK, United States and South Korea who helped me materially and psychologically. All of my PhD colleagues at the Bartlett School of Architecture and their brilliant research drove me to pursue my own research ardently. I thank the UCL Urban Laboratory for their generous support of my independent yet related research during my PhD programme, without which I would not have been able to make smooth progress in my study. Jay Chun in New York lent his professional talent to improve the overall quality of most photographs, and Jungah Hong in Seoul supported me with drawing articulate diagrams. Diana M. Linton has been with me throughout the writing process, helping me with English editing and stylising. Without her shared understanding of the research and sophisticated writing skills, I would not have been able to convey my ideas as they are now. I am also grateful to Suki Kim and Jay Lee for their consistent support from across the globe. Their views and comments on some of the detailed parts of this research helped me to clarify my ideas and gain more confidence in my own thoughts. Professor Sang-kyu Kim has enduringly supported me, helping me access Korean academic materials from abroad. Sky Kim's constant psychological support helped me to have confidence in doing what I am doing. Last but not least, I am more than grateful to my parents for their patience and unflinching support for my study.

Introduction

Landfill in the study of urban ecology[1]

In Anglo-American architectural history and theory, ecology has risen as a matter of concern since global warming became a worldwide issue in the latter half of the twentieth century. During the post-war era, English architectural theorist Reyner Banham (*Architecture of the Well-Tempered Environment*, 1969, *Los Angeles: The Architecture of Four Ecologies*, 1971) examined the social and urban interactions of the crucial constituents of the city, albeit his technology-based utopian vision for the built environment was distant from the relational design approach toward human co-existence with nature. Scottish landscape architect and urban planner Ian McHarg (*Design with Nature*, 1969) analysed ecosystems, providing a more environmentally responsible approach to the built environment. In the following decades, architectural and urban planning experts focused on the city's 'sustainable ecosystems' (Richard Rogers and Philip Gumuchdjian, *Cities for a Small Planet*, 1997; Herbert Girardet, *Creating Sustainable Cities*, 1999) and emphasised the productivity of green spaces, particularly that of public parks in the city.[2] This was particularly true after the UN announced the 1987 Brundtland Report, which addressed the 'three pillars' of sustainable global development (human, environment and economic sustainability).

Especially since the turn of the twenty-first century, both architectural theorists and practitioners have dealt with the subject of the landfill and its regeneration—most notably, transforming it into a public park—as the primary topic of research in association with the ecological discourse. Architectural theorists have illuminated the indissoluble relationship between nature and the built environment as a part of the study of ecology during the 1990s and 2000s (e.g. Richard Ingersoll in *The SAGE Handbook of Architectural History and Theory*, 2012).[3] Architectural academics have also expanded the boundaries of the field by integrating architecture with the urban, which encompasses the social dynamism of urban space and takes both landscape urbanism and ecological urbanism into consideration (Mohsen Mostafavi and Ciro Najle eds., *Landscape Urbanism*, 2003; Charles Waldheim ed. *The Landscape Urbanism Reader*, 2006; Mohsen Mostafavi and Gareth

Doherty eds., *Ecological Urbanism*, 2010). In the ecological discourse on architecture and landscape architecture, the transformation of the New York Fresh Kills Landfill (1948–1997)[4] into Freshkills Park (2010–present) has been influential (James Corner, *The Landscape Imagination*, 2014). Amongst others, James Corner, the landscape architect of Freshkills Park, presents the idea of the 'field' as a 'space-time ecology' that considers all urban forces and agents as continuous networks of inter-relationships. In light of this, James Corner attempts to distinguish landscape urbanism or ecological urbanism, from landscape architecture, which has evolved from McHarg's work, because it fails to address the socio-cultural and political environment.[5]

This study investigates 1) the meaning of Nanjido Landfill (1978–1992, Seoul), as an inhabited unsanitary landfill operating in a developing East Asian nation throughout its industrial era, and 2) the meaning of Nanjido Post-Landfill Park[6] (2002–present), as a park regenerated under globalised neoliberal circumstances, from the urban ecological perspective. It also explores 3) how the disciplines of architectural and urban history and theory make sense of urban ecological studies on the landfill, especially the material and immaterial characteristics of the waste.

This study's examination of the inhabited Nanjido Landfill and its transformation into Nanjido Post-Landfill Park, from the urban ecological perspective and within the larger frame of Seoul City's urban redevelopment, has two main objectives. First, by exploring how 'the relational' dynamics between environmental and social ecologies determine the urban ecology of a site and vice versa, this research will extend the ecological approach in architectural and urban studies from a sustainable design method for continuous human-centred development to an ecological explanation of the habitat and inhabiting. Therefore, I suggest that we think of Nanjido Landfill and Nanjido Post-Landfill Park's construction not only as a sanitation measure and/or environmentally conscious regeneration of urban space, but also as a part of the relational phenomena of the urban ecology. This perspective will enable international audiences to view the spectacular technology-driven urban re/development plans, significant in the cities of developing countries, as a part of the relational urban ecology.

Second, as this study is the first English-language account of the case study on a landfill and landfill-turned-park of South Korea, a relatively lucrative developing nation in Asia, it widens the scope of landfill research in ecological discourses, while defining it as a global issue. In so doing, this study will provide a new approach to understanding the material (environmental) and immaterial (social) levels of waste management as symptoms of ecological dis/equilibrium in the context of global socio-economic systems and planetary urbanisation.

Since the chronology of South Korea's political and economic shifts demonstrates the ways in which social ecology changes the value of the landscape in Seoul's urban space, this study includes a historical analysis to examine the urban ecology of the landfill and park site. In the study of a landfill's creation and transformation, in particular, the historical method shows how the accumulation of variegated land usage and human occupation render an ecological structure. It is also a way of fabricating a diachronic study of the environment's occupation with a synchronic examination of the site.

The ecological stance of this study is rooted in the relational aspect of Félix Guattari's three registers of ecology—the environmental, the social and the human subjectivity (Félix Guattari, *The Three Ecologies*, 2000 [1989]). This research uses these conceptual instruments to redirect the ecological approach in architecture toward an epistemological approach in which ecological equilibrium is principal (Rachel Carson, *Silent Spring*, 2000 [1962]; Lorraine Code, *Ecological Thinking*, 2006; Peg Rawes ed. *Relational Architectural Ecologies*, 2013). In this regard, ecology coincides with ecofeminism (Verena Conley, *Ecopolitics*, 1997). In other words, we can view the inhabited landfill as a social structure within capitalist urban spaces as well as an environmental one. Modern waste management systems that sequester 'waste' or the unclean or the inappropriate are social constructs (Mary Douglas, *Purity and Danger*, 2002 [1966]; Zygmunt Bauman, 'Dream of Purity', 1995 and *Wasted Lives*, 2012 [2004]). For this urban ecological study of Nanjido Post-Landfill Park, I draw on concepts of 'the urban', as more recently articulated by Andy Merrifield (Andy Merrifield, *Henri Lefebvre*, 2006, *The Politics of the Encounter*, 2013 [2012] and *The New Urban Questions*, 2014). Through these research processes, this study will suggest a relational urban ecology as a revisioned approach to architectural history and theory in an era of globalised capitalist urbanism.

Landfill and landfill regeneration[7]

Landfills are waste disposal sites and management facilities where organic materials decompose under bio-chemical processes, releasing a mixture of synthetic materials that pollute the natural environment. Landfills have been the subject of concern in diverse disciplines, including geology, civil engineering, environmental studies, public health and ecology, each with different understandings and interpretations. Geology and earth science define landfill as land people construct for refuse,[8] but civil engineering and environmental conservation see it as the disposal of waste material by burying it underground.[9] The concept of the landfill in public health deals with the risks toxic refuse imposes on the ecosystem and the economic life of garbage collectors, amongst other issues.[10] Ecology and environmental management—biological and physical engineering—investigate elements that carry environmental risks, including leachate and hazardous gas.[11] This study defines the landfill as a man-made landscape created by burying the waste of industrial urbanisation. It also acknowledges that landfills integrate waste into the natural environment where human beings may reside and/or work.

The history of landfills dates back to ancient civilisations (e.g. The Roman Empire), but the history of the large-scale modern landfill parallels that of industrialisation and urbanisation. As the urban population significantly increased throughout the mid-eighteenth and nineteenth centuries in the industrialised societies of the West, waste treatment and disposal became one of the major concerns for the maintenance of an appropriate environment.[12] After WWII ended, municipal landfills were established both in the West (Europe and the United States) and in other parts of the world. Unfortunately, by the latter half of the twentieth century, most were unsanitary landfills, created by large-scale dumping of an untreated mixture of waste. There were few official landfill systems in place in Asian nations before the mid-twentieth century since industrialisation and urbanisation had begun over a century later in Asia than in the West. However, the time gap between the West and developing regions shrunk rapidly after WWII because new economic industrial states became globalised and the consumerist social phenomena gathered momentum. As a result, even though most landfills of developing countries, particularly those of Asia, opened decades after those in the West, both the global North and South confronted similar problems of overflowing unsanitary landfills, particularly since the late 1980s and 1990s.

In western landscape architecture and urban studies, researchers have mainly dealt with the subject of landfills as part of strategies for the reclamation designs of closed landfills, which would transform former landfill sites into recreational leisure facilities and heighten the region's economic value. In recent examples, contributions of advanced technologies to solving issues of waste management in the neoliberal capitalism are significant. One of the most prominent cases of reclamation is the former Fresh Kills Landfill in New York, which opened as Freshkills Park around 2010.[13] James Corner's Field Operations designed this exemplar landscape architecture project during the new neoliberal capitalist era. Based on Corner's understanding of urban ecology, Freshkills Park is a large-scale landscape park that accommodates diverse leisure facilities to serve the general public of New York (Mohsen Mostafavi and Ciro Najle eds., *Landscape Urbanism*, 2003; Charles Waldheim ed., *The Landscape Urbanism Reader*, 2006; Caroline Klein et al. eds., *Regenerative Infrastructures: Freshkills Park, NYC*, 2013; James Corner, *The Landscape Imagination*, 2014).

In Asia, Japan applied the landfill reclamation method as early as 1981 (Aoba Elementary School, former Hachida Landfill, 1968–1973, reclaimed since 1981). However, Japanese landfill reclamation is not currently the most common response to a landfill closure because of the financial constraints and potential environmental risks. Only some of Japan's closed landfills have been reclaimed as parks, sports arenas, parking spaces, advanced waste treatment facilities and civic farms (Imazu Sports Park, former Imazu Landfill, 1975–1999, reclaimed since 1992; Moerenuma Park, former Moerenuma Landfill, 1979–1990, reclaimed since 2005).[14] In Taiwan, former landfills were transformed into the Fudeken Remediation Park (former Fudeken Landfill, 1985–1994, reclaimed since 2004) and the Kaohsiung Metropolitan Park (former Shichinpu Landfill, 1977–1999, reclaimed since 1999). These public parks have art galleries, sports arenas and power plants that collect the methane gas generated from the former landfills, providing power for approximately 7,000 households over 10 years.[15] In Singapore, where most waste is treated by incineration, the Semakau Landfill (1999–present), an offshore sanitary landfill built on an island, is an exceptional case as it is simultaneously used as a landfill and recreational site (opened in 2005).[16] In Australia, the Sydney Olympic Park (opened in 2000) was established on ten closed landfills, which is one of the most significant environmental legacies of the Sydney 2000 Olympic Games.[17]

In South Korea, Seoul City witnessed its first historical landfill reclamation in May 2002, just before the 2002 FIFA World Cup Korea/Japan. The former Nanjido Landfill, the major municipal landfill in west Seoul (Sangam-dong), was transformed into the Nanjido Post-Landfill Park. With financial support from the City, it provides leisure facilities, including an artist residence and campsite. Each of these landfill regeneration cases is distinct for its particular regional socio-economic circumstances. With this in mind, we can surmise that Nanjido Landfill's reclamation history parallels those of Taiwanese and Japanese landfill practices. Also, given that the Nanjido Post-Landfill Park plans coincided with an international sporting event, it is also analogous to the Sydney Olympic Park. In this case, we can say that large-scale international events and urban regeneration may have brought about Nanjido Landfill's reclamation. Nanjido Landfill's transformation was also part of the Sangam-dong area's urban redevelopment, which included building the Digital Media City, one of the leading smart cities in the nation. Lastly, in the context of the urban regeneration of the global economy, the landfill reclamation unfolded in a climate vigilant of humankind's environmental responsibility.

Waste management research

Research on the issues of unsanitary landfills and their environmental hazards has become a subject of concern for the late twentieth and early twenty-first centuries. Accidents related to pollution have led to specialised research and policies on landfill management on the bureaucratic level. One example is the discovery of drums of toxic cyanide dumped in a children's playground in the UK in the late 1970s.[18] Meanwhile, leachate and toxic vapour leaked into housing at the Love Canal site in New York in 1977,[19] and polychlorinated biphenyls (PCBs) seeped into rice oil in Japan in 1968. All of these decisive accidents alerted the public to the severity of the pollution in unsanitary landfills. In response, the West began to conduct research on the problems of hazardous waste in the landfills in the 1970s. National or municipal governments mainly commissioned these efforts, and administrations subsequently established related policies. During the 1980s, governments monitored the policies initiated during the 1970s and continued to develop new approaches to waste treatment. In the United States, the Resource, Conservation and Recovery Act of 1976, the Solid Waste Disposal Act Amendments 1980 and the Hazardous and Solid Waste Amendment of 1984 developed many of the initial rules and regulations. Meanwhile, the UK focused on domestic and industrial waste management methods, including the development of efficient incinerators.[20] In the 1990s, research on leachate and gas emissions and on industrial and commercial waste was developed to find physical and bio-chemical solutions to the pollution from overflowing landfills (Roy Harrison and Ron Hester, *Waste Treatment and Disposal*, 1995; Paul Williams, *Waste Treatment and Disposal*, 2005 [1998]), and systematic regulatory policies were established both nationally and internationally.[21] Inquiries into landfill closures and post-landfill management along with studies on the safer maintenance of existing landfills proceeded simultaneously during this decade. Landfill research notably increased worldwide at the turn of the twenty-first century and continued to increase in developing countries, particularly in the Asia-Pacific region (Chinese studies by Yu-Chi Weng et al., 'Management of Landfill Reclamation with Regard to Biodiversity Preservation, Global Warming Mitigation and Landfill Mining', 2015).[22]

Since the 1990s, the hazardous risks of unsanitary landfills have likewise been major subjects of concern in the fields of civil engineering and environmental science. However, these fields have not devoted as much attention to studies of the interrelationship between the landfill (land), as both a natural and artificial environment, and human beings (society). The research on the human impact of landfill sites in the discipline of architectural history has also seen less progress. Instead, architects have approached it as a structural and system-based urban investigation by way of sustainable design, for example, that employs organic materials and energy-saving technologies. Alternatively, while reflecting on the dichotomy between natural and built environments in modernism, some architectural researchers and practitioners in these fields have viewed environmental matters as a part of the urban fabric (Rem Koolhaas and Edgar Cleijne, *Lagos*, 2007). In the 2000s, the field of architecture expanded the boundary of sustainable design toward a broader social ecology by integrating political, social, anthropological and philosophical studies (Richard Ingersoll, 'The Ecology Question', 2013[23]; Peg Rawes ed., *Relational Architectural Ecologies*, 2013). In the same way, some scholars have engaged with dirt and ruins as constituents of the holistic urban environment (David Gissen, *Subnature*, 2009; Brian Dillon, *Ruin Lust*, 2014).[24]

The interest in land and its relationship with the habitat emerged in the field of art, dating back to the 1960s and 1970s; for example, Robert Smithson examined the relational planetary system of the earth, human beings and materials in exchanges with minimalist and conceptual artists.[25]

Other active artists concerned with land at that time include Nancy Holt, Ana Mendieta, Patricia Johanson and Mel Chin.[26] Artists who have specifically engaged with the landfill in relation to critiques of the consumerist economy, material culture and anthropocentrism and their environmental impacts are Agnes Denes (*Wheatfield: A Confrontation*, 1982), Mierle Laderman Ukeles (*Touch Sanitation*, 1979; *Flow City*, 1983)[27] and Hilary Powell (*Deconstructing Demolition*, 2015). French filmmaker Agnès Varda's *The Gleaners and I* (2000) cast the role of waste in food production and Lucy Walker and Karen Harley's film *Waste Land* (2010) illustrated the social and community value in the inhabited landfill.

In South Korea, Nanjido Landfill research has been conducted since its opening in 1978 and throughout the 1980s when it was still active. Four articles were published, discussing the site's environmental degradation, living conditions and the government's post-landfill plans.[28] Since the landfill's closure to the present, a great deal of geological and bio-chemical research has been conducted; during the planning period for the landfill's stabilisation and regeneration (1993–1997), 39 research papers were published as sources of support for the reclamation of the closed landfill.[29] Based on these studies, Seoul City published *A Preliminary Plan for Environmental Protection and the Nanjido Landfill's Stabilisation* (1992), *A Preliminary Report on the Long-term Land Use Plan for Nanjido Landfill* (1992) and *A Plan for the Nanjido Landfill Stabilisation Construction* (1996). Fifty-nine monitoring reports on the geological and bio-chemical changes of the former landfill site were published during the Landfill's regeneration and park construction period (1998–2001). Since Nanjido Post-Landfill Park's opening (2002–present), Seoul City and environmental organisations have continued to publish monitoring reports.[30] In contrast to the existing studies' focus on technical contributions to Nanjido Post-Landfill Park's construction, this research describes the transformation of Nanjido Landfill into Nanjido Post-Landfill Park from the urban ecological perspective, concentrating on the notion of waste (material and social) and its treatment.

Research methodologies

Historical approach to the landfill study

This study employs humanities-based methodologies ranging from those of architectural and urban history, landscape architecture, art history and aesthetics. Through this multidisciplinary approach, it attempts to identify the characteristics of Nanjido that lie beyond the scope of environmental engineering and science discourses. In terms of its method, the research specifically follows the historical development of post-war South Korea with the nation's political-economic shifts at its centre. The landfill may be a naturally and artificially generated topographical entity, but it is also the historical outcome of human activities. Of course, these aspects are inextricably entwined, so this study focuses on how the history of the site's inhabitation has transformed its topography and re-formulated its identity. The historical lens aims to evaluate the transformations of the site, from wetland/farmland to a landfill and to a park, in the nation or city's political and socio-economic context.

In order to view the site's accumulated history, this study applies the concept of theorised history, which philosophers and socio-political scholars had developed. Edward Casey, Michel Foucault, Pierre Bourdieu and ecological theorists Lorraine Code and Verena Conley discuss the socio-political meanings of a place and regard the diachronicity or historicity of a site as more important than other factors. Their ideas on temporality in the study of space support the importance of examining history to understand the site as a diachronically sedimented entity. While discussing eighteenth-century spatiality, Foucault asserts that 'space itself has a history'[31] in his essay titled 'Of Other Spaces' (1967). Following Foucault's insistence on the genealogy of space, Casey suggests viewing space as something that is variable throughout history in *The Fate of Place* (1998).[32] We can also find the concept of *habitus* in Code and Conley's definition of ecology; their interpretation adds a diachronic temporal plane to the synchronic spatial plane in the concept of ecology. Using Bourdieu's definition of *habitus* (as the notion of an embodied history), Code affirms that this concept entails the historical sedimentation of a community's cultural experiences.[33] Likewise, when we view the study of the urban ecology as inclusive of interactions between the environmental and social, it carries historicity within its spatial concept of habitat.

A historical examination is not only necessary to demonstrate the visible topographical changes of the site, but also to delineate the invisible socio-spatial characteristics inherited from preceding decades. On the surface, the landscape of the Nanjido region has dramatically changed in its form and function. Under the surface, however, its marginalised image has persisted throughout the modern history of South Korea. That is, the site was appropriated for controversial purposes throughout its history as a centre for juvenile reform or a war orphanage during its wetland/farmland era and then as a waste treatment facility during its landfill period. Even after Nanjido Post-Landfill Park covered this 'different' past in the name of environmental awareness, the layers of its accumulated history cannot disappear. They remain an essential component of the place. In short, a historical investigation into the site enables us to understand the creation and subsequent reconfiguration of Nanjido Landfill, and the environmental and social impacts of these transformations.

Field work and primary sources

The information obtained from diverse forms of primary sources has established the main structure of this study. Sources range from printed materials and visual resources on the Nanjido region and landfill, to interviews with

19 individuals who have lived and/or worked on the site. The archival materials include academic research, government publications, mass media reports and non-fiction novels, and visual resources consisted of maps, photographs, TV documentaries and so forth. The 19 interviewees are in some way related to the site's ongoing use and contemporary status. Garnered from these verbal and visual materials, details about the Landfill's political, socio-cultural and natural characteristics and about its transformation into the Park form the cornerstone of this research and help to establish its intellectual focus.

I found a large number of printed materials through the National Assembly Library of Korea, the National Library of Korea and Seoul Library. The major archives encapsulate Seoul City's urban plans for the Nanjido region. Through these libraries, I also found a few academic studies on the living conditions in Nanjido Landfill during its early stages and on Sangam-dong's regional development. My research additionally included newspaper reports on Nanjido and Seoul City's waste management and sanitary control policies (e.g. the use of DDT) since the 1950s to the present.

My visual sources complement the printed materials by reinforcing our understanding of Seoul City's sanitary management during its extensive industrialisation and urbanisation. The images show the changing landscape of the Nanjido region, the scale of Nanjido Landfill, the environmental conditions of the landfill residents' living and working climate and the transformed landscape of the Post-Landfill Park. In addition to the archival photographs of the City and those taken by me personally (e.g. current views of Nanjido Post-Landfill Park), several artists allowed me to use their photographs of the landfill period and views of the closed landfill. I also refer to several videos produced by the major TV networks of South Korea.

Sometimes supporting or modifying the journalists' views, the interviews deepened my understanding of the site. I selected 19 interviewees from different fields so that different voices from different positions could be heard. They also provided insight into diverse aspects of the Landfill and its transformation into the Post-Landfill Park as well as into their different perspectives. Regarding the interview processes, I obtained potential interviewees' contact information from their employers, and then, contacted them via emails or phone calls to invite them to speak with me. I conducted the interviews as hour-long one-on-one conversations. For individual interviews, I prepared questionnaires based on objective facts as documented or reported in existing publications, being careful to avoid any leading questions that might compromise the interviewee's intentions.

Theoretical background

Waste management and ecological equilibrium

The relation between waste and sanitation in this research draws from anthropologist Mary Douglas's concept of 'dirt' (*Purity and Danger*, 1966) and sociologist Zygmunt Bauman's notions of 'dirt', 'waste' and 'surplus'. According to Douglas, no object *is* 'waste' by its intrinsic qualities, but some objects are categorised as waste under the ideal of modernisation. Also, 'eliminating', Douglas notes, 'is not a negative movement, but a positive effort to organise the environment'.[34] Inheriting this idea, Bauman claims that an object becomes or receives the identity of waste by human design. In his argument, design refers to the intentional practice of modernisation, which is rooted in the creation of order through the separation of the appropriate (inclusion) from the inappropriate (exclusion). Bauman explains the modern man's obsession with order as a rule of modern society; within its constraints, mankind separates normality from pathology, health from illness, the accepted from the rejected and what counts from what does not. Through this observation, Bauman delineates what waste means and how society suppresses, hides and controls it.

Using this line of thought, this study explores how Seoul City had controlled material and human waste (e.g. the sick who are considered dirty and abnormal) to maintain the cleanliness of the urban space during the nation's high industrial era. Also, Bauman extended Douglas's idea of 'dirt'—'any matter out of place'—to the broader concept of 'waste' in 'Dream of Purity' (1995) before fully expanding the meaning of 'waste' in the context of a globalised economic system in *Wasted Lives* (2012 [2004]). Bauman's notion of 'waste', both in the material and immaterial social sense, explains the issue of managing garbage as well as marginalised populations (e.g. unemployed, immigrants outcasted from their own [national] territory or refugees) in the global context. This notion also functions as a touchstone throughout this research. In this sense, we can view the landfill as an opportunity to explore the relational dynamics between the material (environmental) and immaterial (social) layers of waste.

Amongst the recurring terms in this research, such as 'dirt', 'garbage', 'refuse', 'waste' and 'surplus', I distinguish the meanings of each term and apply them to a number of different contexts. First, I use the term 'dirt' as it was originally defined by Douglas and developed by Bauman—the most symbolic abstract sense of 'the improper' under modern criteria. Second, I use the two terms 'garbage' and 'refuse' interchangeably to indicate material objects that have been discarded. They have the intense social connotation of denigration; in other words, the objects or people are so dirty and/ or useless that we must separate them from other groups, which we accept as appropriate to the norms of the modern city. Third, I also use Bauman's conception for my definition of 'waste', which indicates any matter including humans that is somehow inappropriate to the modern industrial society and recent era of global economy. Fourth, 'surplus'[35] is the explicitly economic term that Bauman developed and used largely in the context of global economic systems to emphasise that economic profitability and a human being's capacity for production and consumption determine the value of an object.[36] Premised on his notion of 'liquid modern'[37] and consumerist culture, Bauman further develops the idea of waste with respect to the 'surplus' of global capitalism to be a 'surplus' of both material objects and human beings. In the historical context of South Korea, Seoul City used the idea of 'surplus' as a socio-economic tool to manage inappropriate entities and often make them invisible in the urban space. In other words, I use the term 'surplus' throughout this research of the Nanjido region's history when the focus of the landfill and sanitation-related discussion turns from the political or ideological to the economic. In this sense, we can conceive the Nanjido Landfill's transformation into Nanjido Post-Landfill Park as a matter of 'surplus' management with regard to the new millennium's concept of environmentalism.

Bauman further argues that inequality, originating in the imperialist era, allowed the more affluent, modern parts of the globe to find *global* solutions to their *locally* produced waste problems. He then asserts that under the current globalised socio-economic circumstances, each locality has to seek *local* solutions to *globally* produced problems.[38] Most significantly, material waste and unemployed populations, consequences of industry, have dramatically increased under the global economic system, and each nation struggles to resolve these problems on a local level. Overall, national or corporate entities that operate on a global scale are largely responsible for these global problems. As a result, the recent tendency to address these problems is a modified version of old methods of waste management; for example, the affluent West/North continues to impose the burden of overflowing hazardous material waste on regions outside its own territory, the so-called global South. When we chart the flow of material waste, it is clear that these global phenomena most severely damage the environmental habitats of the least affluent and most powerless regions, and this has ultimately led to the vicious cycle of ecological disequilibrium.

Meanwhile, inhabited landfills—where garbage collectors reside while manually recycling materials in and around the landfill site—are more common in developing countries than in relatively advanced ones. What's more, these less affluent nations import more environmentally hazardous waste from developed regions. The recycling business in developing regions is generally relegated to the low-income class, and garbage collectors who work manually in landfills are exposed to more fatal health risks. In this global economic system, the waste management system and business of South Korea lie somewhere in-between the advanced and less developed regions. On the one hand, relatively wealthier nations, including South Korea, export hazardous digital waste to underdeveloped countries. On the other hand, in the more affluent nations, since corporate systems had taken over the waste treatment business in the late 1990s, the manual job of waste management in the local waste market is still more marginalised in the sense that incomes are much lower than garbage collectors' in the landfills of prior decades. Therefore, we must view changes in the waste management systems and the role of the landfill in the context of the global economy, the worldwide flowchart of waste and its relationship with ecological equilibrium between the global North and South. A deeper grasp of the global economy's background will give us a more comprehensive understanding of the opening and closure of Nanjido Landfill and its transformation into Nanjido Post-Landfill Park.

Ecology as the way of inhabiting the world

Looking back at the history of ecology, one of the early Darwinians, Ernst Haeckel (1834–1919) coined the term 'ecology' in *Generelle Morphologie der Organismen* (1866). The term is rooted in the word 'eco' (*oikos*), and Haeckel defines it as 'the study of household of nature', or relationships between organisms and their environment.[39] The idea that Nanjido Landfill and Nanjido Post-Landfill Park are a part of the relational ecologies of the environment, social dynamics and human subjectivity forms the fundamental premise for this study. Thus, the ecological and eco-political theories (Félix Guattari, *The Three Ecologies*, 2000 [1989]; Lorraine Code, *Ecological Thinking*, 2006; Rachel Carson, *Silent Spring*, 2000 [1962]; Verena Conley, *Ecopolitics*, 1997; Peg Rawes ed., *Relational Architectural Ecologies*, 2013) underlie the discussions in this research. Examining the Nanjido region while referencing ecological theories enables us to view its multiple historical faces—pre-landfill (wetland), landfill and post-landfill (park) periods—as the totality of a habitat that has a relationally interconnected environment and society. Secondly, it helps us to see the site's complexity as a way of inhabiting the world.

This study takes Guattari's ideas on ecology, mainly developed in his *The Three Ecologies* (2000 [1989]),[40] as one of the primary discourses of relational ecologies. His arguments about multivalent ecologies—environmental, social and mental ecologies[41]—based on globalised economic and social phenomena, efficiently explain the relational aspects of the environment and social ecology. In this study, I contend that the Nanjido region's transformations from 1) wetlands to 2) a landfill and to 3) a park are the result of the relational dynamism between the environmental and social ecologies, and that the transformed urban spaces continue to generate new relational ecological dynamism.

Amongst the key ideas of ecological thinking and practices explored in *The Three Ecologies*, Guattari develops the concept 'ecosophy', an ethico-political articulation of the three ecological registers (the environmental, social and mental ecologies of human subjectivity). He suggests that it is a potential way to address the political-economic and social phenomena of Integrated World Capitalism (IWC)[42]; considering that the 1980s was the beginning of the global economy and the period that laid the groundwork for neoliberalism, we can regard the IWC as an advanced capitalist economy satisfying the primary conditions for globalised neoliberal capitalism. Guattari's exploration of the environmental and social problems of the

IWC informs us of the socio-economic changes that occurred on a global level. These changes grew in significance during the 1980s in the West and persisted into the 1990s and early twenty-first century. The same phenomena also arose almost concurrently in the developing parts of the world. As one of the developing countries or latecomers, South Korea, though slowly transforming into a neoliberal economy in the 1980s and the early 1990s, faced a great shift toward neoliberal capitalism in the late 1990s. Therefore, we can apply Guattari's argument surrounding the IWC to the local situation in South Korea at the turn of the new millennium.

Historically, the 1968 movements were the decisive moments when the relational ecology gained importance. For example, feminist and environmental politics drew attention to the links between the affluent West/North white male-centred world order and environmental degradation. Conley reads the events of 1968 in France as a turning point in ecological awareness insisting that, in the post-1968 era, 'ecology takes on the double meaning of being at once a natural and social concern aimed at measuring habitability'.[43] While arguing for a shift toward a horizontal worldview, Conley and Code contend that ecology and feminism commonly aimed for improved habitability and planetary equilibrium.[44] Despite the different focus on ecology and women's insights within the feminist movements (deep ecology and ecofeminism[45]), they largely share a commitment toward reshaping the worldview by revising modes of engagement with the environment, society and human subjectivity. Ecological thinking suggests a 'relational' alternative that creates horizontal relations, and proposes a relationship between humankind and the environment, or the human and non-human. With this in mind, Donna Haraway argues, 'we must find another relationship to nature besides reification, possession, appropriation and nostalgia'.[46] She suggests a systemic shift from an anthropocentric point of view to a materialist or geological worldview that invites us to reconsider dominant subject–object relationships and our overall existential conditions. In the study of urban landscape, a revised understanding of the relationship between the environment, social relations and human subjectivity can lead us to an alternative conception of the urban landscape. For instance, we could interpret it as an ecological practice that produces humankind and allows for inhabitation, instead of as a design for the backdrop of human residence.

The study of Nanjido Post-Landfill Park draws on ecological thinking (Lorraine Code), ecopolitics (Verena Conley) and relational ecologies (Peg Rawes), as the research ultimately seeks an alternative approach to understanding South Korea's ecology, which has been confined to the preservation of the natural environment. Here, the discursive perspective is not limited to a scientific epistemology of ecology but is concerned with ecofeminist discussions and practices that reflect upon human engagement with the urban environment, and the production of social relations and human subjectivity.

Decades prior to these movements, in *Silent Spring* (1962), Rachel Carson showed an exemplary case for the practice of ecological research as ecological knowledge production and the researcher's position as a situated subjectivity. For Carson, ecology means the study of 'patterns in nature' and the examination of how they operate and interact with each other.[47] Carson's study on toxic chemicals (e.g. DDT) and their fatal damage to nature, and ultimately the ecosystem, is meaningful in ecological studies, particularly because her research methods, position and rhetoric differ from those of institution-based scientists at that time. Carson bases her production of knowledge on extensive, lengthy field experience, so it is, to borrow Haraway's term, 'situated knowledge'[48] that prioritises the matter of concern over the matter of fact.[49] Carson's ethical and situated approach prompts us to explore the value of ecology for what Code calls an epistemic responsibility.

In this research, Carson's practice for the prohibition of chemical insecticides is the foundation for the discussion on the nationwide practice of sanitisation and use of DDT in South Korea as part of the study of environmental and social ecologies. Regarding the national/municipal sanitary policies, it highlights the active use of DDT in the post-war developing country of South Korea and considers the United States' influence on the use of pesticides for fumigation after WWII. A historical understanding of DDT fumigation and its ban in the post-war space of Seoul allows us to perceive the global dimensions of environmental concern. This study further argues that Nanjido Landfill, partly because of this socio-political background, functioned not only as a facility for the material cleanliness of the city but also as a symbolic urban structure that represents the notion of sanitation in the urban space.

To examine the meaning of ecology as habitat and the manner of inhabiting, it is necessary to revisit the root of the term; etymologically, *oikos* means household and ecology pertains to making habitats where human beings live in accordance with the environment. Code defines ecology as follows:

> [Ecology] is a study of habitats both physical and social where people endeavour to live well together; of ways of knowing that foster or thwart such living; and thus of the *ethos* and *habitus* enacted in the knowledge and actions, customs, and social structures, and creative-regulative principles by which people strive or fail to achieve this multiple realisable end (Italics added).[50]

We can see here that the philosophical conceptions of *ethos* and *habitus* form the cornerstone of ecology. *Ethos* in Greek originally means 'accustomed place', 'custom' and 'habit', which is equivalent to the Latin *mores*. It also forms the root of *ethikos*, meaning 'showing moral character'. Its ideas have expanded to indicate human and non-human behaviour in the environment and to map human beings' relations to the physical and social environment. Deleuze develops the idea of ethology, defining it as 'the capacities for affecting and being affected that characterise each thing'.[51] In the sense that any being only exists through reciprocal influences within a network of relations, ecology becomes a matter of and conditioned by our sociabilities.

Adepts like Code and Conley use the concept of *habitus* in an attempt to set up a diachronic axis in ecological studies. Bourdieu defines *habitus* as the notion of an embodied history, internalised as second nature, and thus, forgotten as history. *Habitus* is also an 'active presence' that deals with the sense of place constituted by the sedimentation of experiences a human being carries in relation to the power structure that shapes such places.[52] As such, the notion of *habitus* adds a temporal dimension to our understanding of habitat, so we perceive it as more than just a physical space; we recognise it as a space of individual and collective histories and memories. The diachronic aspect of ecology shapes the structural frame of this research, and thus, the study elucidates how the environmental and social factors of the Nanjido region have accumulated over the decades.

Ecological thinking concerns not only the habitat *per se*, but also how humankind inhabits the world, or how human beings relate to their habitat. Location is not a simple background upon which things exist, but is constitutive of the inhabiting processes. Conley suggests that ecological thinking offers a better way of inhabiting the world, and, adding to this, Code defines 'inhabiting' as an active, thoughtful practice socially, emotively and responsibly engaged.[53] From here, ecological thinking as an epistemic project—to be aware of the conditions of justly inhabiting the world—moves toward living in the physical and social location: the embodied locatedness. Drawing on the idea of *habitus*, this study investigates the Nanjido region through the lens of the embodied history and active presence of the site. It additionally examines how the site and its people affect and are affected by different entities within the urban context of Seoul, especially concerning Nanjido's landfill period.[54]

The urban[55] in the globalised capitalist urbanism

As notions of the world city or global city have advanced in relation to the changes in global economic systems, the term 'the urban' has widely spread, simultaneously enhancing and specifying its meanings. This study takes Andy Merrifield's recent discourse on 'the urban' as its primary theoretical basis because Merrifield, drawing on Henri Lefebvre's discourse on 'the city' and 'the urban', develops these concepts in relation to the digital technology-based network society and globalised neoliberal economic system. Since Merrifield is a thorough scholar of Lefebvre, the Lefebvrian ideas of 'the city' and 'the urban' as well as early discussions on planetary urbanisation surface in his interpretations of the previous generation.[56]

Merrifield relies on Lefebvre's urban theory, particularly his definition of the urban: 'It's not in space that people act: *people become space by acting*. They *are* space' (Italics in original). Accordingly, he prioritises social relations grounded on 'common notions',[57] and even equalises the network and urban societies.[58] Further developing this idea, Merrifield states that, 'In an invironment[59] the "performance" itself engineers and creates the spatial relations as well as the behaviour of every participant'. In this respect, the Lefebvrian idea of the urban resembles ideas valued by ecological thinking, such as sociabilities, communities, the relational and the way we inhabit the world. That is, urban and ecological thinking may share substantial ideas concerning the ethical and political aspects of an individual's claims to his/her own place and to that of others. 'The urban'[60] discourse affirms that neither the separation of fragments and contents nor their confused union can define the urban phenomenon. It incorporates a *total reading*, combining the vocabularies of geographers, demographers, economists, sociologists, semiologists and others. 'The urban' is neither an integrated totality nor a separated facet, outside of the formal frame of the city. In this vein, studying the Nanjido site is about not only the form or formation of the city, but also the *total reading* of the relational dynamism of the area's political, economic and social attributes, which render the city of Seoul.

The concept of 'the urban', in this study, fundamentally comes from globalised capitalist urbanisation and its establishment of a planetary 'fabric' or 'web' of urbanised spaces. As globalised capitalism emerged and networks link societies of different geographical regions, the term 'planetary urbanisation' has become significant, especially after the regional economic crises in East and Southeast Asia in 1997–1998, Russia in 1998, Argentina in 2001 and so on, culminating in the global crash of 2008.[61] It is highly likely that the interrelated socio-economic conditions including waste management and global ecology amongst regions are more explicit now than ever before.[62] In his discourses, Merrifield has never thought of 'the urban' as separate from capital, democracy or class issues. David Harvey's notions also materialise in Merrifield's development of the concept within discussions of globalised capitalist urbanism. In line with Lefebvre's conception of the urban, Harvey prioritises the urban as a socio-political and economic outcome, particularly the relation between economy and class, and its impact on the city's formation.[63] Regarding the relationship between capitalism and urbanisation, it is worth referring to the twenty-first-century concept of 'surplus', which Harvey addresses.

> Capitalism is perpetually producing the surplus product that urbanisation requires. The reverse relation also holds. Capitalism needs urbanisation to absorb the surplus products it perpetually produces. In this way, an inner connection emerges between the development of capitalism and urbanisation.[64]

What is notable about his argument is that, by relating urbanisation to the issue of class, Harvey questions the authority in command of the connection between urbanisation and surplus production and use.

In the study of the urban ecology of Nanjido Landfill and Nanjido Post-Landfill Park, 'the urban' refers to the political and socio-economic conditions and their relational dynamism, which perpetually shifts, generates and, as a result, forms the environment of the city. Throughout the latter half of the twentieth century, South Korea had experienced radical shifts in its political and economic systems. Meanwhile, global capitalism became the predominant condition, determining the urban situations of all regions since the 1980s onwards (globally and in South Korea, since the 1990s, particularly after 1997). Therefore, we can regard the study of 'the urban' as a critique of the structural urban problems caused by the changing political circumstances and the capitalist system. The problems of urbanisation essentially focus on the potential for emancipatory politics, or the 'possible urban worlds'[65] that are systemically suppressed by power relations and institutional arrangements. To clarify the definition of 'the urban' in the study of Nanjido Landfill and Nanjido Post-Landfill Park, it is necessary to understand how South Korea's shifting political and economic/corporate powers influence the site's socio-economic value.

A bordered space

The Nanjido Landfill community's work and working environment, which dealt with material refuse, made the community members outcasts of the modern norms of both urban and architectural sanitation.[66] Fundamentally, dominant modern socio-economic value systems pointedly stigmatise such groups of people as outcasts. This research (particularly Chapter 3) focuses on the community lives of the Nanjido Landfill period, and discusses how the new socio-economic system creates an invisible border between different social groups. For the study of the Nanjido Landfill community, understanding the notion of the outcast in the late twentieth and early twenty-first centuries is therefore necessary. In Bauman's discussion, the outcast, or the surplus population, includes political refugees and asylum seekers, mainly flowing from former colonised regions into former colonising countries. These individuals remain in a separate zone or suspended territory between the inside and outside of the urban space. The concept of the outcast shares its foundation with Giorgio Agamben's notion of *homo sacer*, or 'bare life' in front of a sovereign power,[67] which says that economically wasted entities and the politically abandoned beings share common ground.[68] This study on Nanjido Landfill sees the landfill as a colonised and suspended space within its own national/municipal territory and the garbage collectors as 'non-citizens' who are 'legally' citizens but not treated as proper citizens under certain socio-economic conditions. As unemployed or temporary workers, landfill workers are readily abandoned, especially under neoliberal economic systems, so frustration against the national/municipal power has increased within this population. Taking a step further, we can link the landfill space to the discourse on camp—the 'intermezzo' space consisting of flows that deconstruct or transform the positions of related subjects. Drawing on Agamben (*Homo Sacer*, 1998), and Deleuze and Guattari (*A Thousand Plateaus*, 1987), sociologist Bülent Diken describes the characteristics of camp as follows:

> Every camp (as machine) consists of flows selected and transformed in the 'intermezzo' into an inside and outside. The camp and the related subject positions are thus coded, decoded and recoded by a differential system (for example, class, sex, crime record, age, consumption patterns and so forth) that can be further decomposed into more diversified differential systems […] when the camp becomes a space of negotiation and proliferation of heterogeneous inside and outside, it can constitute a 'city'.[69]

Viewing the inhabited landfill as a space in-between camp and city, in this sense, will permit alternative interpretations of the landfill populations' socio-political positions on the border and their potential sociability.

Since Nanjido Landfill was a separated zone of presumably impoverished inhabitants, we can compare it to the ghetto, a marginalised inner-city space. The ghetto refers to an area of urban poverty mainly occupied by racial, ethnic minorities, often immigrants who are in enduring, if not perpetuating, poverty. Loïc Wacquant, regarding the upheaval in Los Angeles in 1984, argues that the label 'race riot' hides another deeper phenomenon, claiming that there is 'a class logic pushing the impoverished fractions of the working class to rise up against *economic deprivation and widen social inequalities*' (Italics in original).[70] He contends that such incidents are the results of socio-economic changes that have led to the polarisation of the class structure. In his assessment, Wacquant deems massive unemployment and labour precariousness (the socio-economic), relegation of populations to decaying neighbourhoods (the environmental) and heightened stigmatisation (the mental) responsible for the violence 'from above'.[71] Especially when destitution persists, the ghetto becomes stigmatised and segregated from surrounding regions because of the overarching criteria for sanitation. The birth of the Nanjido Landfill community is different from the creation of a ghetto, generally a neighbourhood of people in an ethno-racial low-income class. However, we can still regard it as a ghetto in the sense that the landfill community was composed of a population of individuals with precarious socio-economic positions,[72] lacking readily marketable skills or assets, and the landfill site was segregated from the other parts of the city. Regardless of the physical proximity to the city centre, Nanjido Landfill remained insulated for issues of sanitation. Even since its transformation into Nanjido Post-Landfill Park, it has been relatively disconnected from the neighbouring regions.

One of the most symbolic elements related to the stigmatisation of Nanjido Landfill and its habitants was odour. Historian, Alain Corbin (*The Foul and the Fragrant*, 1994), relates smell, especially stench, to class and occupation, suggesting that extremely toxic odours impregnate the bodies of certain occupational groups in the low-income working class. This study on Nanjido Landfill sees odour as a decisively crucial factor in three ways: first, as Corbin argues, odour identifies a person's job and class; second, odour formulates an invisible border between a site (the landfill) and other regions of the city; and third, odour crosses the invisible border it creates. The contradictory dual aspects of smell to make and cross borders simultaneously are the most significant characteristics of the landfill and its community. By crossing the very border that it creates, it threatens the authority of the society's existing norms. Furthermore, this study extends the discussion of Nanjido Landfill's isolation, either by the natural frontier (e.g. The Han River stream) or by its ghettoised imagery, to the sense of unease, or the sense of place or placelessness of Nanjido Post-Landfill Park. The last chapter in this research surveys site-specific artistic projects by previous resident artists of the Park's SeMA Nanji Residency in an attempt to answer the question of the site's unease. Examining their art pieces demonstrates how the artists use the site's materials and sensory aspects, particularly the olfactory and auditory senses, to represent the site's decades of conflicting historical layers. These aesthetic projects explore the extraordinary sense of uneasiness perceived in the site and the reasons for the community's lack of emotional attachment to the current Park. The analyses of the artistic practices on Nanjido Post-Landfill Park reinforce our consideration of the relational ecological structure amongst the natural environment, social relations and human subjectivity. Understanding the ecology of the site, in turn, evokes the sense of place or the awareness of its concealment.

Notes

1 Matthew Gandy insists on the term ecological urbanism instead of urban ecology due to his concern that urban ecology restricts urban issues within the boundaries of biological studies. He argues that it remains a dubious part of multidisciplinary studies since it is unable to configure its own historiography. For this very reason, however, this research takes 'urban ecology' as leaving interdisciplinary boundaries open (Matthew Gandy, 'From Urban Ecology to Ecological Urbanism', *Area*, Vol. 47, No. 2, 2015: 150–154). In this article, Gandy reminds us that ecological urbanism shares the conceptual agenda of landscape urbanism, which calls for a synthesis between landscape and urban design (Frederick Steiner, 'Landscape Ecological Urbanism', *Landscape and Urban Planning*, Vol. 100, No. 4, 2011: 333–337; Charles Waldheim, 'Weak Work' in Mostafavi and Doherty eds., 2010: 114–121).
2 Peg Rawes ed., *Relational Architectural Ecologies*, London: Routledge, 2013: 4–7. For the historical development of the discourse on architectural ecologies, see the introduction to this publication.

3 Richard Ingersoll in *The SAGE Handbook of Architectural History and Theory*, C. Greig Crysler et al. eds., 2012: 29.
4 Robert Moses and the City of New York established Fresh Kills Landfill in 1948. In 2003, the landscape architecture firm Field Operations (James Corner) was selected as the planning and design consultant for Freshkills Park.
5 James Corner, 'Terra Fluxus', *The Landscape Urbanism Reader*, Charles Waldheim ed., New York: Princeton Architectural Press, 2006: 29–30.
6 The official name of the park is the World Cup Park. This research, however, calls it Nanjido Post-Landfill Park to stress the material co-existence of the land-fill mass and the park instead of the Park's association with a national/municipal sporting event.
7 In this study, the term 'landfill regeneration' incorporates 'landfill rehabilitation' and 'landfill reclamation'.
8 *A Dictionary of Geology and Earth Sciences*, Oxford: Oxford University Press, 2015.
9 *A Dictionary of Construction, Surveying and Civil Engineering*, Oxford: Oxford University Press, 2013.
10 *A Dictionary of Environment and Conservation*, Oxford: Oxford University Press, 2012; *A Dictionary of Public Health*, Oxford: Oxford University Press, 2007.
11 *The Encyclopaedia of Ecology and Environmental Management*, Oxford: Blackwell Publishing, 2009.
12 Throughout the latter half of the nineteenth century in the UK, a series of Nuisance Removal and Disease Prevention Acts were introduced. In the United States, the Corporation of Georgetown Washington DC announced a law pro-hibiting waste disposal in public spaces in 1795. By 1856, Washington had a citywide waste collection system supported by taxes. See Paul Williams, *Waste Treatment and Disposal* (Chichester: Wiley, 2005 [1998]); H. A. Neal et al., *Solid Waste Management and the Environment* (NJ: Prentice Hall, 1987); and E.A. McBean et al., *Solid Waste Landfill Engineering and Design* (NJ: Prentice Hall, 1995).
13 For other landfill regeneration cases in Canada (Keele Valley) and Israel (Hiriya Landfill), compared with Freshkills Park, see Benjamin Lawson, 'Garbage Mountains', PhD Dissertation in History, The University of Iowa, 2015.
14 Yu-Chi Weng et al., 'Management of Landfill Reclamation with Regard to Biodiversity Preservation, Global Warming Mitigation and Landfill Mining', *Journal of Cleaner Production*, Vol. 104, 1 October 2015: 367; Yu-Chi Weng et al., 'Proposal of an Integrated Evaluation Approach on Final Disposal Sites with Regard to Future Reclamation', *Journal of Japan Society*, 2013: 313–320.
15 Weng et al. 2015: 367–368; Veolia Environment, S.A., *Report of Methane Treatment in Fudeken Sanitary Landfill*, Environmental Protection Bureau, Kaohsiung City Government, Taiwan, 2014.
16 Regarding the technical matters and ethical concerns about the offshore Semakau Landfill, see Jeffrey Kok Hui Chan, 'The Ethics of Working with Wicked Urban Waste Problems', *Landscape and Urban Planning*, Vol. 154, October 2016: 123–131.
17 Weng et al., 2015: 368; E. Laginestra et al., 'Towards Sustainable Urban Ecology, the Olympic Site Message', *Geographical Education*, Vol. 14, 2001: 27–30.
18 Government-supported research to monitor the waste has continued since the late 1970s. See Paul Williams, 2005: 3; J. Sumner, *Co-operative Programme of Research on the Behaviour of Hazardous Wastes in Landfill Sites—final report*, The Policy Review Committee for the Department of Environment, UK, London: H.M.S.O, 1978.

19 Elizabeth D. Blum's *Love Canal Revisited* (KS: University Press of Kansas, 2008) re-illuminated the Love Canal incident from the perspective of social relations.

20 Paul Williams, 2005: 3–5.

21 The National Waste Strategy is a requirement of all member states of the EU (Paul Williams, 2005: 5).

22 In 2015 and 2016, international journals published an extensive volume of bio-chemical research on Chinese landfills and waste disposal. Consideration must also be given to the influence of environment-related events or movements, such as the UN Climate Summit (2014) and the UN Climate Change Conference (2015).

23 *Journal of Architectural Education*, Vol. 45, No. 2, 2013: 125–127.

24 Esther da Costa Meyer argues that architectural history has struggled to generate approaches that respond to new urban conditions (Esther da Costa Meyer, 'Architectural History in the Anthropocene', *The Journal of Architecture*, Vol. 21, No. 8, 2016: 1203–1225).

25 For more on Smithson's land art, see Lynne Cooke et al. eds., *Robert Smithson* (Berkeley and London: University of California Press, 2005) and Ann Reynolds, *Robert Smithson* (Cambridge, MA and London: MIT Press, 2003).

26 For artistic projects on land, see Barbara Matilsky, *Fragile Ecologies* (New York: Rizzoli International, 1992).

27 Ukeles declared the 'Manifesto for Maintenance Art' in 1969. She had been an unsalaried Artist in Residence of the New York City Department of Sanitation for decades since 1977. In 1990, she became directly involved with the regeneration of Fresh Kills Landfill. For Ukeles's full projects and ideas, see Kari Conte ed., *Mierle Laderman Ukeles* (Amsterdam: Kunstverein Publishing, 2015).

28 Amongst the four works of literature published during the active landfill period, two are socio-anthropological studies on the poverty of the Nanjido people and the environmental concern. The other two are preliminary studies on the post-landfill's regeneration and the park development plan. Socio-anthropological studies include Chae-Sung Chung's 'The Poverty of Nanjido Residents as its Social Relations', *Korean Society for Cultural Anthropology* 21, 1989 and Yeok-Ki Kim and Se-Hoon Jang's 'Internal Differentiation and Reproduction of Poverty', *Korean Sociological Association* 21, April 1988. Research on social inequality increased in the late 1980s partly because the abiding democratic movement overthrew the military dictatorship in June 1989 and the closing of the landfill was approaching.

29 There were three studies on the social and environmental subjects of the site during this period. For social studies on Nanjido Landfill at these times, see Ho Lee, 'The Residence Right of the Nanjido Residents', *City and Poverty*, Vol. 21, March 1996: 47–67.

30 Social studies on the conflicts between the city's urban redevelopment plan and Nanjido Landfill residents were published during this period. Seo-young Park's 'Historical Transformations of Sangam-dong Based on the Lived Experience of Its Local Residents' (Master's thesis in Cultural Studies at Yeonsei University, 2004) is an ethnographic study on the lives of North Sangam-dong residents. Seung Won Chi in 'Self-Identity and the Principles of Justice of Small Groups in a Community' (*Korean Social Theories*, Vol. 40, Fall/Winter 2011) attempts a strictly theoretical analysis by applying sociological methodologies to the Nanjido Landfill community. Jieun Shin's 'Waste, City and the Reconstruction of Urban Consumption' (*A Journal of Contemporary Social Science of Korea*, No. 17, 2013: 17–38) reviews the symbolic meanings of waste in consumer culture.

31 Foucault asserts that it is not possible to disregard the intersection of time with space (Michel Foucault, 'Of Other Spaces', trans. by J. Miskowiec, *Diacritics*, Vol. 16, No. 1, Spring 1986: 22).

32 'Not only is space not absolute and place not permanent, but the conception of each is subject to the most extensive historical vicissitudes. The extremity we now enter is that of the historicity of our subject: a challenging prospect indeed' (Edward Casey, *The Fate of Place*, CA and London: University of California Press, 1998: 297–298).

33 Lorraine Code, *Ecological Thinking*, Oxford: Oxford University Press, 2006: 28; Pierre Bourdieu, *Logic of Practice*, Cambridge: Polity Press, 1992 [*Le sens pratique*, Paris, 1980]: 53, 56. Although the structuralist and poststructuralist spatial understandings of history oppose the linear progressive development of the grand narrative, they rarely deny the accumulated temporality of the historical space.

34 Mary Douglas, *Purity and Danger*, London: Penguin, 2002 [1966]: 12.

35 The concept of 'surplus' in landscape is also based on the utilitarian ideal of the modern industrial society; it connotes that these spaces are supplementary and unnecessary for functional purposes, thereby valueless. Transformations of these spaces into areas for leisure, entertainment, art and green (parks) purposes— instilling cultural and environmental values—reflect the creation of exchange value in addition to the use value of the spaces. To avoid confusion with the existing notion of 'surplus space' in landscape architecture, which is often confined to the idea of abandoned sites, or ruins, this study uses the notion of 'surplus' in a fundamentally economic way, particularly in the globalised neoliberal context.

36 This is originally from Marx's theory of capital. Harvey, in *Rebel Cities*, accounts for the relationship between the concept of surplus in the capitalist industrialisation and urbanisation (David Harvey, *Rebel Cities*, New York: Verso, 2013 [2012]: 5, 136–138).

37 Bauman argued that modernity at once signified a dual nature, ordering and rationalising tendencies, and the constant change and overthrowing of tradition, as Karl Marx phrased 'all that is solid melts into air'. Later, drawing on this idea, he began to use the term 'liquid' as opposed to 'solid', representing mobility and change. He then described the transition from modernity to postmodernity as one from solid modernity to liquid modernity. See Zygmunt Bauman, *Liquid Modernity*, Cambridge: Polity Press, 2000.

38 Zygmunt Bauman, *Wasted Lives*, MA: Polity Press, 2012 [2004]: 5–6.

39 Peg Rawes ed.: 1. Sharon Kingsland also noted that Haeckel defined the new science of *ökologie* as the study of households (Sharon Kingsland, *Modelling Nature*, IL: University of Chicago Press, 1995 [1985]: 11).

40 In response to the upsurge in environmental concern, the original French version was published in 1989, 3 years after the 1986 Chernobyl nuclear power plant accident.

41 Conley analyses that, amongst Guattari's three ecological registers (environmental, social and mental), the subject's mental ecology is the most important because it includes myriad relations, from which we draw diagrams that contribute to the construction of ever-changing ecosystems (Verena Conley, *Ecopolitics*, London: Routledge, 1997: 96).

42 For Guattari's discussion about Integrated World Capitalism (IWC), see Félix Guattari, *The Three Ecologies*, trans. by Ian Pindar and Paul Sutton, London and New York: Continuum, 2000 [1989]: 21, 32–35.

43 Conley: 26, 110.

44 The term *ecofeminism* itself originates from French feminist Françoise d'Eaubonne's book *Le feminisme ou la mort* (1974). In her book, D'Eaubonne urged that ecofeminism would contribute to remaking a new model for the planet (Code, 2006: 16–17; D'Eaubonne, 'Time for Ecofeminism', trans. by

Ruth Hottell in *Ecology*, Carolyn Merchant ed., NJ: Humanities Press, 1994: 176).

45 Of ecological theories, deep ecology, represented by Carolyn Merchant (*The Death of Nature*, 1980) and Val Plumwood (*Feminism and the Mastery of Nature*, 1993), exposes the covert political implications of anthropocentric science's domination over nature. Meanwhile, scholars such as Luc Ferry (*The New Ecological Order*, 1995) criticise post-1968 ecological movements, particularly deep ecology and ecofeminism, for their fascist tendency that may threaten humanism, democracy and enlightenment.

46 Donna Haraway, 'Otherworldly Conversations, Terran Topics, Local Terms', *Science as Culture*, Vol. 3, No. 1, 1992: 65.

47 Carson's approach meshes well with Sharon Kingsland's conception of ecology, defined as 'the study of patterns in nature, of how these patterns came to be, how they change in space in time, why some are more fragile than others'. Furthermore, population ecology studies 'how populations interact with the environment and how these interactions give rise to the larger patterns of communities and ecosystems [...]' (Code, 2006: 26; Sharon Kingsland, *Modelling Nature*, IL: University of Chicago Press, 1995: 1).

48 Haraway, 1992: 70, 92.

49 Carson elucidated the environmental hazard of disseminating synthetic pesticides through quantitative research results based on her fieldwork. She intended for people to see nature as an integrated, organic and living whole. See Linda Lear, Afterword to Rachel Carson, *Silent Spring*, London and New York: Penguin Classics, 2000 [1962]: 258–264.

50 Code, 2006: 25. Conley mentions ethos and structuralist Bourdieu's concept of habitus (Conley, 1997: 94, 96) and their relationships to ecopolitics. She argues that structuralism opened our minds to the environment and showed how the health of the human sensorium depends on its empathy for the wealth of nature in which it lives. Without the Lévi-Straussian ethnography, she insists, many of the poststructural works evincing a new awareness of the woman and the environment would not be what they are (Conley, 1997: 147–149). Conley also asserts that in Guattari's writings, the main hypotheses of Lévi-Strauss's ecology are mobilised, updated and diverted for active, pragmatic politics with a complicated relation to ecofeminism (Conley, 1997: 92, 102).

51 Gilles Deleuze, *Spinoza*, trans. by Robert Hurley, CA: City Lights Books, 1988 [1970]: 125–126.

52 Bourdieu: 53, 56.

53 Conley, 1997: 114.

54 It is in line with Deleuze and Guattari's concept of *milieu*. In Deleuze and Guattari's definition, *milieu* is 'the site, habitat, or medium of ecological interaction and encounter'. As such, it emphasises the continuities of the interrelationships between humans and non-humans and the cross-connections amongst the physical, biological, chemical, social, ethical and political (Gilles Deleuze and Félix Guattari, *What is Philosophy?* trans. by Graham Burchill, London and New York: Verso, 1994 [1991]: 192–196).

55 When Lefebvre used the term 'the urban' in 'The Urban Revolution' (1970), he stated that 'the urban' is an abbreviated form of 'urban society'. He demonstrated 'I use the term "urban society" to refer to the society that results from industrialisation [...] Instead of the term "post-industrial society"—the society that is born of industrialisation and succeeds it—I will use "urban society", a term that refers to tendencies, orientations and virtualities, rather than any preordained reality'. He also declared that we could only define urban society as global or planetary, whereas the urban is locally intensified and does not

exist without that localisation. See Henri Lefebvre, 'The Urban Revolution' in *The Global Cities Reader* (1970), Neil Brenner and Roger Keil eds., New York: Routledge, 2006: 408.

56 Merrifield points out that Lefebvre's notion of 'the city' had morphed into 'the urban' between 'The Right to the City' (1968) and 'The Urban Revolution' (1970). In 'The Urban Revolution', Lefebvre explored concepts like 'planetary', 'urban practice', 'non-city', 'anti-city', 'implosion-explosion' and 'centre of power' (command-and-control), ideas mentioned in the discussion of the urban or globalism. Lefebvre argues against the idea that we can halt large-scale urbanisation—consolidation of the world city—through revolutionary praxis from rural peripheries. Versions of this notion continue to reappear in debates on postcolonialism and in Michael Hardt and Antonio Negri's influential musings on the 'multitude'. See Michael Hardt and Antonio Negri, *Multitude* (New York: Penguin, 2004) and Editor's introduction to Lefebvre, 'The Urban Revolution' (1970) in *The Global Cities Reader*, 2006: 407–408.

57 Andy Merrifield, *The Politics of the Encounter*, GA: University of Georgia Press, 2013 [2012]: 47, 122, 130–131. 'Common notion' is originally the Spinozian sense that he evinced in Part Two of *The Ethics* (1677). Robert Abraham argues 'no one can ever hope to get to the core of Spinoza's thought without a perfect understanding of what is, I should say, *the* nuclear theme of Part Two, namely *adequate thinking of common notions*, Spinoza's Second Mode of Cognition and to him the basis of all reasoning [...] If two sides are to advance jointly, they have to agree on the conditions laid down in the underlying definitions and axioms. In all such cases, mutual understanding is the prerequisite of mutual progress' (Italics in original) (Robert Abraham, 'Spinoza's Concept of Common Notions', *Revue Internationale de Philosophie*, Vol. 31, No. 119/120, 1/2, SPINOZA, 1977: 27–28).

58 Merrifield, 2013: 70.

59 Merrifield, 2013: 124. 'Invironment' is the term that pertains to the environment of information.

60 It originated from Lefebvre's 'The Urban Revolution' (1970).

61 Regarding the urban roots of capitalist crises, see Harvey, *Rebel Cities*, 2013 [2012], chapter 2.

62 Neil Brenner and Christian Schmid, amongst other scholars, are known to have claimed 'planetary urbanism'. In a recent declaration, however, Brenner and Schmid clarified their argument and position on the term 'planetary' as follows: 'We have never claimed it to describe the epistemology of the urban. Our work is focused on the problematic of urbanisation, capitalist urbanisation in particular. [...] We have constantly emphasised the constitutively uneven, variegated nature of capitalist urbanisation, including its planetary configuration' (Neil Brenner, 'Debating Planetary Urbanization', Working Paper, Urban Theory Lab, Harvard GSD, Summer 2017. See also Christian Schmid, 'Planetary Urbanization', *Critique of Urbanization*, Neil Brenner ed., Basel: Birkhäuser Verlag, 2017: 186–191).

63 When claiming the right to the city (RTTC), Harvey determined that urban issues are inseparable from capitalism, and re-emphasised Lefebvre's argument of four decades prior that a contemporary revolution must be urban (Harvey, 2013 [2012]: 25). See also David Harvey, *Justice, Nature and the Geography of Difference*, MA: Blackwell Publishers, 1999 [1996].

64 Harvey, 2013: 5.

65 Harvey, 1999.

66 As for people's assimilation to their job based on smell, see Alain Corbin, *The Foul and the Fragrant*, London: Picador, 1994.

67 Bülent Diken and Carsten Laustsen call this 'zone of indistinction'. See Bülent Diken and Carsten Laustsen, 'Zones of Indistinction', *Space and Culture*, Vol. 5, No. 3, 2002: 290–307. For related discussions, see Bülent Diken, 'From Refugee Camps to Gated Communities', *Citizenship Studies*, Vol. 8, No. 1, March 2004: 83–106 and Loïc Wacquant, *Urban Outcasts*, Cambridge: Polity Press, 2008 [2007]: 229–256.

68 Conley cites Étienne Balibar's argument that an eco-subject must be a citizen-subject and have minimum existential territory (citizens do not necessarily have to be nationals). See Verena Conley, 'The Ecological Relation', *Relational Architectural Ecologies*, Peg Rawes ed., 2013: 283; Étienne Balibar, *Droit de Cité* (Paris: Les editions de l'aube, 1998).

69 Diken, 2004: 104; Gilles Deleuze and Félix Guattari, *A Thousand Plateaus*, MN: University of Minnesota Press, 1987: 432.

70 Wacquant, 2008: 22.

71 Ibid.: 24–25. It is not unlike the proclamations of the defenders of Marxist theories of societal transformation. Immanuel Wallerstein claims that class formation is supposed to wash away ethnicity and create a global class structure (Immanuel Wallerstein, *Historical Capitalism with Capitalist Civilization*, London and New York: Verso, 2011 [1983]: 73–93).

72 Regarding the precarity of habitation, see Judith Butler, *Precarious Life*, London: Verso, 2004.

1   Transformations of Nanjido

Throughout the late twentieth century, after Korea's liberation from Japanese colonial rule (1910–1945) and the Korean War (1950–1953), South Korea, especially the capital city of Seoul, went through highly rapid urban reconstruction and change, and the Nanjido region[1] (Figures 1.1, 1.2) in the western part of Seoul, in particular, experienced dramatic transformations in its function and landscape.

We can divide the history of Nanjido into three different phases. First, after the Korean War until the early 1970s, Nanjido is generally recorded as an underdeveloped yet idyllic wetland/farmland. Second, from the mid-1970s to the mid-1990s, the entire site of Nanjido was used as Seoul's Municipal Solid Waste Landfill,[2] the largest unprecedented landfill in the city thus far. Third, after Nanjido Landfill's closure, the landfill site was rehabilitated throughout the late 1990s when Nanjido was reclaimed as a landscape park called Nanjido Post-Landfill Park (World Cup Park), its opening coinciding with the 2002 FIFA World Cup Korea/Japan (Figures 1.3, 1.4, 1.5).

Chapter 1 examines the environmental and social characteristics of Nanjido during these three different phases. It also studies how two instances of historical transformation transpired in the context of the nation's shifting political economy and in relation to Seoul City's urban planning for the area. By doing so, this study demonstrates how political-economic power engenders the metamorphosis of urban space, re-formulating the landscape and topography and reconstituting the identity of the site (Table 1.1).

In each part of the chapter, I discuss the environmental and social aspects of the region, using key concepts that are representative of each period: 1) nature, 2) waste and 3) regeneration. I then inquire into the specific processes and characteristics of the region's multiple transformations through the lens of South Korea's political-economic systems. Concerning the economy, the pre-landfill, landfill and post-landfill periods notably coincide with South Korea's economic transitions from industrial (1960s–1970s) to post-industrial (late 1970s–1990s) and to neoliberal (1990s–present). Politically, South Korea's longest dictatorial regimes of the 1960s and 1970s attempted to control waste and waste workers in many sites with diverse methods, and opened the large-scale yet unsanitary landfill in Nanjido during their

Figure 1.1 Nanjido region located in Sangam-dong (town) in western Seoul. © Jeong Hye Kim

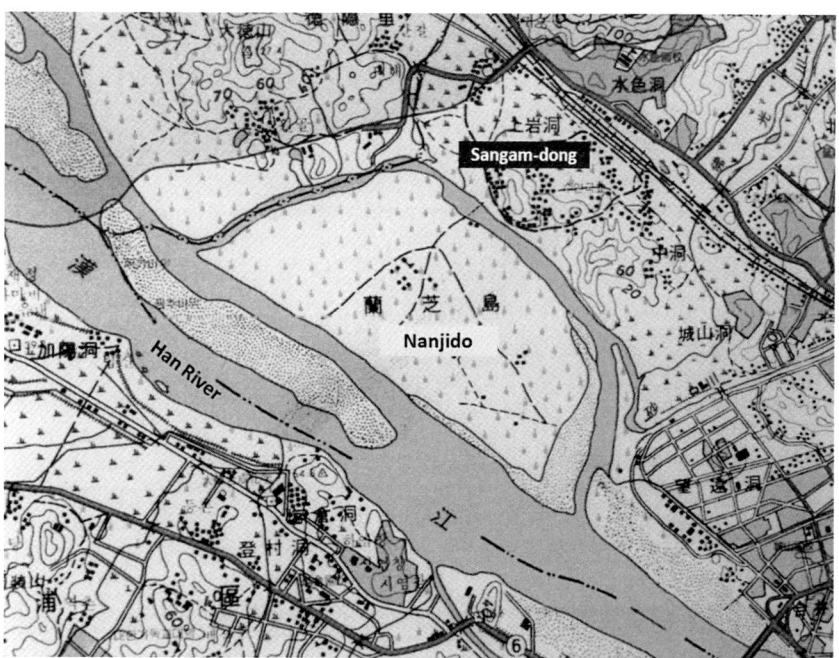

Figure 1.2 Map of Nanjido region. 1974. Map © National Map Museum

Figure 1.3 An afternoon on Saetgang (a small stream of the Han River). The Han River stream surrounded Nanjido during the pre-landfill period. Wetland lined the shore, and locals used boats as the major means of transportation. 1960s. © Jong-Chul Won

Figure 1.4 Nanjido Landfill mound and landfill workers' collective housing complex
from Joon Kim, *Instant Landscape*. 1996. © Joon Kim

despotic rule. A decade later, democratic administrations, which had confronted the globalised neoliberal transformation of the 1990s and the new
millennium, closed Nanjido's overflowing landfill and transformed it into
Nanjido Post-Landfill Park, equipping it with cultural and environmentally
friendly[3] waste treatment facilities (e.g. the SeMA Nanji Residency for artistic activities and the Mapo Resource Recovery Plant for advanced waste
treatment). Analysing the relationship between the political economy and
regional changes shows how political-economic changes not only influence
the landscape but also constitute or damage the equilibrium between environmental and social ecologies. It also explicates the changes, distortions
and strategic uses of concepts like 'nature', 'environment' and 'ecology' in
each era.

Behind the transformations of Nanjido region's landscape and function
lies an enduring characteristic that it has retained for decades: its isolation.
Even though the site is located within Seoul, less than an hour away from
downtown by public transportation, and is legally open to the physical
access of all, the public's[4] negative perception of this region has prevailed
throughout all of its transformations.

First, the major reason for Nanjido's regional isolation is the government's decision to retain Nanjido and its surrounding areas as greenbelt
land. Historically, from the Japanese colonial period until 1963, the government excluded Nanjido from municipal urban planning. In 1964, when the

Figure 1.5 Post-landfill period of Nanjido (currently Nanjido Post-Landfill Park, 2002–present). © Seobu Park & Landscape Management Office, Seoul City

Ministry of Construction of South Korea appointed part of Sangam-dong as a residential area, it restricted Nanjido to a green zone. In the 1970s, despite the City's need to develop more land for economic purposes, Nanjido remained an untouched green zone. As a result, Nanjido's isolation during the pre-landfill period excluded it from industrialisation and urbanisation on the one hand, but preserved the region's natural environment, on the other hand. From 1976 until the early twenty-first century, even during its active landfill period, the site was officially restricted to greenbelt.[5] For this reason, throughout South Korea's most prolific period of industrialisation during the 1960s and early 1970s, the site remained secluded within the urban space of Seoul.

Second, the Seoul City government built public facilities that people generally dislike in Nanjido and its surrounding areas from the Japanese colonial period to this day. These included an orphanage and a Centre for Juvenile Reform. Above all, the City appointed this region as the municipal landfill in 1978, despite the site's official designation as a greenbelt. The landfill period (1978–1992) was decisive in that Nanjido was almost completely segregated from the other spaces of Seoul, including North Sangam-dong.[6] The site's use as a landfill severely degraded its natural environment, ultimately aggravating the region's isolation.

Table 1.1 Transformations of Nanjido and the political economy of South Korea

Nanjido	Year	Politics	Economy
① PRE-LANDFILL PERIOD Centre for Juvenile Reform	19_0–1945	Japanese Colonial Administration	Pre-industrial age (Agricultural industry)
Orphanage / Boys Town (1952/1953–present)	1950–1953	The Korean War	Industrial age
	1961	*5.16 military coup d'état	Dictatorial regimes
	1963–1972	The Third Republic	
	1972–1979	The Fourth Republic	
② LANDFILL PERIOD Nanjido Landfill (1978–1992)	1981–1988	The Fifth Republic **June Democracy Movement (1987)	
	1988–1993	The Sixth Republic	(Transition)
	1993	Civilian Government	Democratic regimes
③ POST-LANDFILL PERIOD	1997	People's Government	***1997/1998: IMF Neoliberal economy
Beginning of Landfill's Stabilisation & Park Landscaping	2002–present	Participatory Government	Post-industrial age
Nanjido Post-Landfill Park (World Cup Park)			

Table by Jeong Hye Kim
*Park Chung-hee staged the 5.16 military coup d'état and took office for nearly 20 years.
**The nationwide movement for democracy generated mass protests in June 1987, which ultimately terminated the decades-long military dictatorship.
***The 1997 IMF financial crisis was the momentum, turning the South Korean economy to a neoliberal system.

Third, Nanjido's most recent transformation from a landfill site into a landscape park was the City's plan to regenerate nature by restoring the degraded site to its pre-landfill state. The City unearthed visuals of the pre-landfill period's natural environment to use as a model for the new era's park. Despite the revitalisation of the natural environment, the socio-cultural value of the site remains questionable, as it is unclear whether Nanjido's isolated characteristic has changed into a more interactive one. We can attribute this doubt to the lingering presence of the site's unpleasant past even after the post-landfill period. Therefore, I assert that there is a perpetually indelible image inscribed on the public's perceptions of the site regardless of the significant landscape transformations. This study also premises that historically consolidated government systems and social dynamics, which can hardly be removed by landscape design alone, have created the site's enduring characteristic—isolation.

Essentially triggered by the political-economic transitions of South Korea, Nanjido's historical transformations are represented by the site's re-formulated landscapes and the coinciding environmental degradation or rehabilitation. Therefore, this chapter accounts for the cross-relationships between South Korea's political economy and Nanjido's environment and landscape re-formulations, exploring how they constituted or damaged the urban ecology of this location.

Pre-landfill period (1945–1977): nature

Appropriation of natural environment

After Korea's liberation and the Korean War, South Korea proceeded with unprecedentedly rapid industrialisation and urbanisation, particularly in the capital city of Seoul, throughout its extended domination under dictatorial regimes in the 1960s and 1970s. During this era, Nanjido remained one of the most underdeveloped towns in Seoul. Because the national/municipal governments owned much of the Nanjido land, Nanjido could preserve its natural environment, or remain in a state of untouched nature, despite the nation's prolific efforts to industrialise. During the Japanese colonial era, the Japanese government owned a large portion of Nanjido's land, and its ownership was transferred to the Seoul City government upon Korea's liberation. Inheriting the land ownership, Seoul City continued the limited use of the site. For example, it restricted regional development by appointing the land solely for public purposes, which led to the construction of facilities deemed unwelcome in central city spaces (e.g. Centre for Juvenile Reform, war orphanage and military training site), and by designating part of the region as a greenbelt zone.

Since the pre-landfill Nanjido region was pre-industrialised and pre-urbanised, most indigenous residents[7] served the agricultural industry, mainly cultivating peanuts and sorghum. Since the river stream and wetland surrounding Nanjido used to be flooded every summer, the site needed infrastructures, such as bridges for transportation and a levee to prevent flooding. However, these were established neither during the pre-landfill period nor during the landfill era. Without bridges, residents would use traditional boats as a means of transportation to cross the wetland and stream and to access the other parts of the city (Figure 1.3). From a developmental point of view, the Nanjido habitants of this period lived a pre-industrial and premodern socio-economic life and were thereby relegated to the lower income classes. From an environmentalist perspective, the forced underdevelopment of the government's greenbelt policy (partly to protect military facilities in the area) eventually preserved the nature of Nanjido.

If we focus on the site's historical subordination to the national/municipal governments for the use of its land for 'anti-social' public facilities, we can view Nanjido as a zone of pollution since it harbours potential threats to the 'clean' environment of the modern city space. Literature published during this period describes aspects of the poor living conditions in the Centre for Juvenile Reform, which mainly housed war orphans, as well as in Nanjido's farms. Several pre-landfill literary essays illustrate Nanjido's pastoral yet labour-laden scenes where homeless or vagrant children and teenagers gathered after the Korean War.[8] Additionally, Samdong Boys Town,[9] one of the largest orphanages, was established in the region during the middle of the Korean War in 1952. According to the news reports published at the time of its opening, the Samdong Boys Town run by the YMCA accommodated vagrant boys, who were mostly homeless war orphans. The news report titled 'Boys Town Opened in Nanjido' (*Kyunghyang Shinmun*, 14 August 1953) describes that the Samdong Boys Town will contribute to cleansing the city space by accommodating and disciplining vagrant homeless war orphans in the new orphanage in Nanjido (The Korean term *burang-a* literally means vagrant, but after the Korean War, it referred to homeless children and teenagers wandering around the city):

> The YMCA has hunted down and gathered the vagrant teenagers to cleanse the city streets of them and opened the Boys Town in Nanjido. It received 150 boys on 12 August at 3 pm. The Boys Town will accommodate and discipline the vagrant boys.

During the post-war era, the Seoul City government prohibited the development of Nanjido Landfill by law, which eventually preserved the region's natural environment and the habitants' pastoral living situation. Meanwhile, by restricting private development, the government and/or public organisations could use Nanjido as a site for the supposedly damaging and uncontrollable members of society in order to create the clean, orderly and safe city space of Seoul. That is, the municipal government appropriated Nanjido to manage potentially threatening urban constituents, thereby protecting the industrialised and modernised urban space of Seoul. In this context, we can contend that Seoul City used Nanjido as a site for socially wasted human beings prior to using it for waste management during its next phase as a landfill.

Idealisation of nature

The underdeveloped environmental conditions provided Nanjido with ambivalent characteristics: it was both a zone under natural or artificial threat and an untouched idyllic green site. In fact, despite the many practical inconveniences of agrarian life of that time, habitants and visitors alike describe this Nanjido of the past as ideally preserved nature of bucolic scenery ideal for leisurely outings. The site's location by the Han River, distant enough but not too far from downtown, practically made the region an ideal site for Seoul citizens to enjoy the natural environment. In this context, even the inconveniences of agrarian lives in the wetland and the lack of infrastructures contribute to furthering the idealised imaginary[10] of the natural environment.

The idealised imaginary of nature in Nanjido of the past resurged upon the site's transformation from the Landfill into Nanjido Post-Landfill Park in 2002, when Seoul City attempted to revive the green environment of Nanjido's pre-landfill period.[11] (It is generally called 'ecological restoration', but the term 'ecology' here is erroneously conflated with a love of nature and environmental-friendliness.) In other words, the idealised imaginary of nature in pre-landfill Nanjido was the symbolic model for the new millennium's environmental revitalisation of Nanjido Post-Landfill Park. Urban imaginaries of the pre-landfill period, likewise, oscillate between a zone of underdeveloped threat and that of idealised nature. Within the modern discourse on civilisation, we generally regard the concepts of 'nature' and 'threat' in the same way that we consider 'the primitive'. Thus, to understand how people perceived Nanjido of the pre-landfill period, we must speculate into two different yet interrelated approaches to nature: 1) the nature–threat discourse and 2) the idealisation of nature.

On the one hand, from the point of view of the nature–threat discourse, the pre-landfill period of Nanjido lacks basic infrastructures, so the prospects for the environment entirely rely on the state of nature and meteorology. In reality, the stream surrounding Nanjido would overflow every summer and bring about recurring floods in the residential area, destroying the human habitat. This perspective implies a proportional relationship between development and wellbeing, and underdevelopment and threat: *development–order–culture–wellbeing* versus *underdevelopment–disorder–nature–threat*. However, a formula that regards 'nature' as something undesigned or underdeveloped for its state of disorder and element of threat embodies the typical developmentalist and anthropocentric point of view. There is virtually little ground upon which we can conceive that a built environment counters the state of nature. Nor can we assert that a built environment guarantees wellbeing by protecting human beings from unpredictable, threatening nature.[12] On the other hand, the building of Nanjido Post-Landfill Park in 2002 was a project intended to revive nature; the revival, here, was grounded on the idealised imaginary of the pre-landfill landscape of Nanjido. On the surface, Seoul City claimed to rehabilitate the natural environment of Nanjido because the landfill period had caused its deterioration. Yet, a psychological yearning for the idealised imaginary of nature, found in the pre-landfill period of the region, lay under the surface. It is an idealised imaginary because, in reality, the pre-landfill Nanjido landscape involved hard labour and inconvenient living conditions as well as nature preserved as a result of its industrial underdevelopment. In *What is Nature?* (1995), Kate Soper explains the lost time-space of 'nature' and the ideological representations of rurality with the idea of 'nostalgia':

> What is at issue here [...] is the post mortem quality of the appeal of 'nature': that it is most forcefully felt only in the wake of its destruction [...] a certain idea of 'nature' becomes more desirable, and the desire for it more manipulable, as the reality it conceptualizes is diminished and degraded.[13]

This environmentalist approach of the new era is premised on a revised formula: *[re]development–order–nature–wellbeing* versus *non-[re]development–disorder–pollution–danger*. What is significant in this new formula is that the 'culture' in the previous formula has been replaced with 'nature' as a condition for 'wellbeing'. By idealising the nature of the pre-landfill era of Nanjido, we have created a model for nature's revival in the post-landfill period. The new era's revitalisation of nature is expected to achieve wellbeing through the design of an orderly state of nature. 'Nature' in this new logic does not refer to the untouched state of nature, but to an artificially built natural landscape. As we need to distinguish practical threats from mythical ones, the same is true of the concept of nature.

In the context of the twenty-first century, the terms 'nature' and 'environment', which are often misconceived and misused as the equivalent of 'ecology' in South Korea, bear positive connotations, and combining them with conceptions of urban regeneration amplifies their benign nuances. As a result, people generally receive the regeneration of nature as an uncontroversial virtue. The dilemma here is that the concept of ecology, which encompasses environmental and social dimensions, is confined to the environmental or even confused with the greenerisation of a landscape, sidelining socio-ecological aspects and the equilibrium between environmental and social ecologies.

In summary, the pre-landfill environment of Nanjido nested in the historical remnants of twentieth-century modern Korea, especially its dark elements. The administrative policies for the region's land use (or restriction of land use) to cleanse the urban space of Seoul were inherited from the colonial era. On the other hand, due to its exclusion from the excessive industrialisation of the 1960s and 1970s, Nanjido could preserve its former natural environment; as environmental-friendliness became a nationwide concern, the new millennium brought the environment's preservation back into the limelight. Nevertheless, given that Nanjido had been used for socially wasted entities and that its risky natural conditions had not been addressed for decades, we cannot conclude that the pre-landfill Nanjido region was in an unadulterated state of unpolluted nature. As for the ecological equilibrium amongst the environmental, social and human subjectivity of Nanjido of those times, its 'untouched' natural conditions, in contrast to its other socio-ecological aspects, have been overemphasised and idealised by contemporary interpretation. With this in mind, we need to view the agricultural industry and natural environment of the pre-landfill Nanjido region not as an underdeveloped or ideal form but as a part of the holistic ecology.[14]

Landfill period (1978–1992): waste

Post-industrial age and landfills of Seoul

This part of the chapter attempts to demonstrate how shifting industrial situations under the ongoing dictatorial regimes and the struggle for South Korea's democratisation influenced the opening and closure of Nanjido Landfill (1978–1992). I examine how the hasty economically driven decision to appoint Nanjido as the municipal landfill site affected environmental degradation in the following decades.

During the postcolonial and post-war years in the late 1940s and 1950s, South Korea experienced serious political confusion; throughout the citizens' struggle for a democratic government, a newly established administration was not systematic enough to fight against corruption, nor firm enough to be independent, so it relied on the United States. During the ongoing conflict between the corrupted government and the citizens who strived for democracy, the military general, Park Chung-hee, staged a coup and assumed office in 1961. Since then, Park's dictatorial regime persisted for about 20 years, and another 10 years of military dictatorship followed his regime. Ironically, it was during these three decades that South Korea accomplished record-breaking economic growth, passing through the industrial and consumerist post-industrial economic systems in a relatively short period.[15] The nation's economic growth rate between the early 1960s and the late 1980s shows the speed of industrial development and urban expansion; the gross domestic product increased by an average of over 8% per year, from 2.7 billion USD in 1962 to 230 billion USD in 1989, reaching over a trillion dollars in 2007.[16] GDP per capita grew from 103.88 USD in 1962 to 5,438.24 USD in 1989, reaching 20,000 USD in 2007.[17] Authoritarian control carried out this motorised economic development by mobilising human and non-human (machinic) resources and launching diverse collective movements in the name of national development, or nationalist patriotism.

As the society's urbanisation developed in concurrence with its industrialisation, a large population gathered in the capital city of Seoul, which eventually made Seoul a megalopolis within approximately three to four decades. Since the 1960s, Seoul City has sustained incessant urban development; this, in the national context, means the modernisation of the urban environment, including the development of built material environments and immaterial social environments, from housing and road construction to social movements for public and private cleanliness. Waste management, in particular, became a crucial issue for the City to address, especially from the late 1970s throughout the 1980s and until the early 1990s when the South Korean economy entered the post-industrial age. As Zygmunt Bauman argues in *Wasted Lives*, capitalist society, especially since the advent of neoliberal capitalism in the post-industrial age, cannot help overflowing with surplus,[18] and thus, the establishment of a separate waste management system is inevitable in or near the city. A landfill is the most fundamental infrastructure, both material and symbolic, required to maintain the appropriate urban environment of post-industrial capitalism. The opening of Nanjido Landfill in 1978 was inextricably intertwined with the nation's capitalist transition from an industrial age to a post-industrial age, which correlated with Seoul City's urban expansion.

• Landfills of Seoul: 1960s–1977

As the scale and method of waste management depend on the amount and type of waste, waste management facilities or landfills change according to shifts in the economic system. Before the opening of Nanjido Landfill, during the 1960s and early and mid-1970s, the military government had managed waste disposal in various places throughout the city space. In the early 1960s, Seoul citizens disposed of waste in the future housing development sites or wetlands because there were no designated landfill sites. It was not until 1964 that the Seoul City government first opened landfills in four towns in Seoul (in Gunja-dong, Sangwolgok-dong, Eungam-dong and Yeomchang-dong). More than a decade later, between 1976 and 1978, the City appointed six more landfills (in Apgujeong-dong, Jangan-dong, Gu-eui-dong, Cheongdam-dong Songjeong-dong and Bangbae-dong) to treat the increased waste (Figure 1.6). While the number of landfill sites increased throughout this industrial age, the types of waste were almost unchanged (biodegradable waste from construction sites, used coal briquettes and mixtures of domestic waste), and so, the method of waste management in these landfills hardly differed from that of unofficial dumping sites.[19]

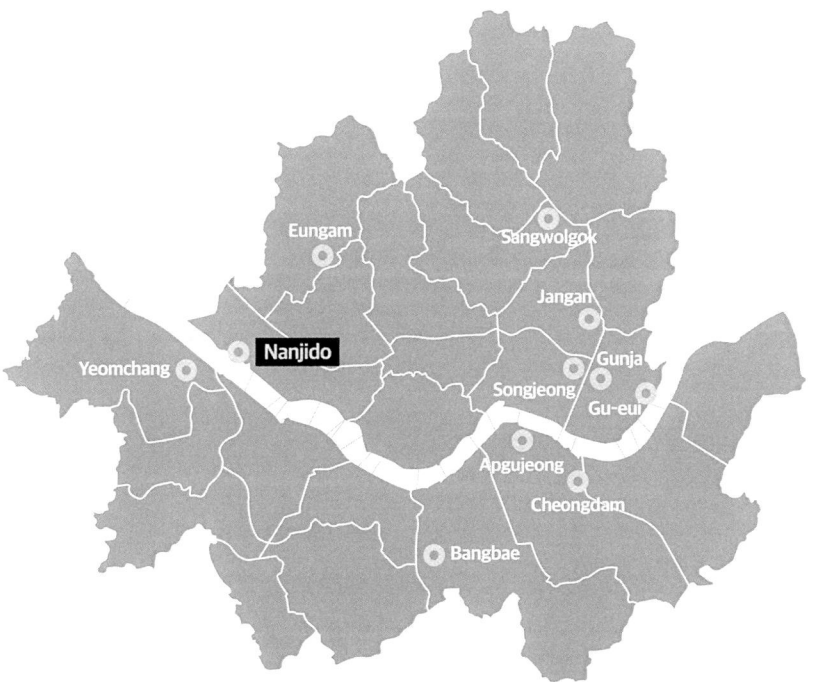

Figure 1.6 Landfills in Seoul, 1960s–1977. © Jeong Hye Kim

- Landfills in Seoul: 1978–1992 (Nanjido Landfill)

Since waste is the direct result of consumption, waste management surfaced as a more crucial issue in the post-industrial era of South Korea, or the high consumerist times from the 1980s until the early 1990s, than it had in the industrial age of the previous decades. In the late 1970s, as South Korea's economic situation changed more rapidly and the nation began to enter a consumerist post-industrial society, most wasted materials changed to non-biodegradable consumer goods that required different methods of waste management, such as composting or recycling. The previous waste disposal methods, such as burying waste under construction sites, were no longer appropriate for managing the massive amounts of waste. Instead, the City needed a discrete area of land that could receive and treat city-wide house-hold waste. Therefore, in 1978, the city government appointed Nanjido as Seoul's official Municipal Solid Waste Landfill, the largest landfill thus far. From March 1978 when the City opened Nanjido Landfill until its official closure in 1992, Nanjido received and managed 77.7% of the Municipal Solid Waste of Seoul[20] (Figure 1.7).

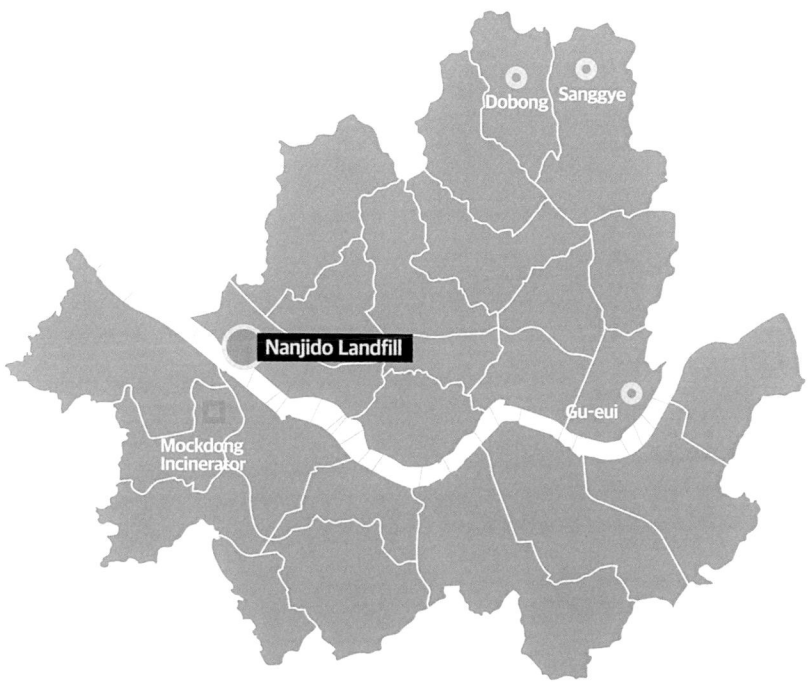

Figure 1.7 Landfills and incinerator in Seoul. 1978–1992. © Jeong Hye Kim

With regard to the political situation, Nanjido Landfill opened during the last period of South Korea's most draconian military dictatorship (the Fourth Republic, 1972–1979), and it closed at the end of the government's transition from a dictatorial regime to a democratic administration (the Sixth Republic, 1988–1993). Regarding the economy, Nanjido Landfill's opening coincided with South Korea's transformation from an industrial to a post-industrial era when the GDP per capita continued to increase and the citizens' consumption simultaneously expanded significantly. Nanjido's active landfill period almost completely overlapped with the peak of South Korea's post-industrial age, converging with its enhanced urbanisation. From the geo-cultural perspective, setting up a Municipal Solid Waste Landfill meant that the city had officially appropriated—or colonised—certain parts of the city for the hygienic maintenance of other parts of the urban space, and, as a result, divided the urban space into waste and non-waste zones.

- Landfills of Seoul: 1992–present

From 1989, as Nanjido Landfill's closure neared, Seoul City planned to build an alternative landfill called 'Sudogwon Landfill' (Capital Area Landfill,

1992–present) outside Seoul to manage the city's waste along with the waste from satellite cities (Figure 1.8). Sudogwon Landfill has been gradually landfilling the waste in all four of its landfills (Landfills 1–4) one after another. The landfill management office built a 36-hole golf course on top of Landfill 1 (1992–2000) after its closure, and it opened a horse-riding course and swimming pool within the Green Bio Complex of the landfill area. After waste disposal filled Landfill 2 (2000–2018), Landfill 3 became the next dumping site, which was expected to reach maximum capacity around 2015. The use of Landfill 4 followed thereafter, and its expected availability was a maximum 50 years. From the onset, the landfill management office has collected gas from the Sudogwon Landfill to produce 1,200,000 kW of electricity per day, which it sold to the Korea Electric Power Corporation. As of 2018, Sudogwon Landfill (especially Landfill 3) announced that it will close in 2025, and thus, each city, including Seoul, has to prepare alternative waste management systems within their own regions by that year.[21]

During the industrial era, towns and communities were responsible for the task of waste control. As the scale of waste became more colossal and the management systems more complicated, waste treatment shifted from

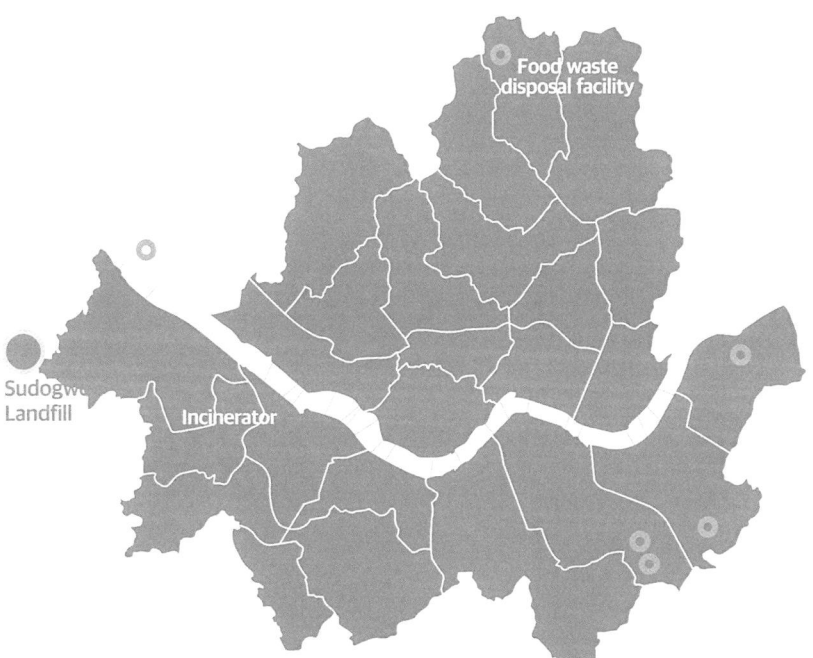

Figure 1.8 Seoul's landfills and incinerators in the Nanjido Post-Landfill period. After Nanjido Landfill had closed, Seoul City appointed Sudogwon Landfill (Capital Area Landfill in the Gyeonggi-do province) as the new Municipal Solid Waste Landfill. 1992–present. © Jeong Hye Kim

public service to private corporatisation. That is, throughout the Nanjido Post-Landfill period, waste management has become a profitable corporate industry operating in the global capitalist economic system; for example, following the logic of the capitalist economy, South Korea now imports lower-priced recyclable waste and exports non-recyclable digital waste, which is often hazardous and toxic, to less developed regions of the world. This exchange of waste has led to new issues of environmental and social ecologies on a global scale.

Environmental conditions of Nanjido Landfill

The selection process of Nanjido as the Municipal Solid Waste Landfill site shows how the national/city government appropriated part of the city space to maintain the cleanliness of the urban space. At the peak of industrialisation, military administrations proceeded with urban development plans as excessively as possible as they did with industrial development, while side-lining the population's ecological rights, particularly those of the economically powerless, and increasing potential damage to the environmental and social ecologies of these groups. One of the main reasons for Seoul City's selection of Nanjido as the Municipal Solid Waste Landfill site was that the City owned a higher portion of its land (Table 1.2). For the national/municipal government, it was relatively easy to take and use its own land for detestable public facilities like a landfill. Even when the nation/city does not own all of the land, the government can budget for the purchase of the rest of the real estate if it holds ownership of a large part of the needed land. In Nanjido's case, the City managed to purchase the private lands in a relatively short period because the regional population was smaller than that in other areas of Seoul. On 3 August 1977, the City declared a plan to use Nanjido as the Municipal Solid Waste Landfill that, they then thought, would receive approximately 10,000 tonnes of waste daily until 1984. Upon

Table 1.2 Land ownership of Sangam-dong including Nanjido and neighbouring town (m², %)

	Total	*Nation-owned*	*City-owned*	*Private*
Sangam-dong	8,331,199	2,887,441	2,889,072	2,554,686
Neighbouring town (Part of Seongsan 2-dong)	704,357	37,999	404,109	262,249
Total (m²)	9,034,556	2,925,440	3,292,181	2,816,935
Total (%)	100	32.4	36.4	31.2

Table by Seoul City

this announcement, the real estate of Nanjido sharply plummeted, and the City was able to purchase the private lands at a reduced rate.[22]

To select the municipal landfill site, the government had to evaluate the candidate site's environmental conditions and check if it met landfill standards (of the United States). In Nanjido's case, however, the City appropriated the land prior to assessing the site's conditions and getting clearance to use it for the landfill. According to Philip Rushbrook and Michael Pugh, technical experts on landfills, there are five major conditions required for landfills: 1) access, 2) natural conditions, 3) land use, 4) population and 5) safety.[23] Nanjido satisfied most of these requisites, including convenient access yet sufficient distance from the airport and populated areas. When it came to safety, however, Nanjido was prone to floods. The flood risk height of a landfill must be less than 13 m, but it was 13.8 m in Nanjido at the time of the landfill selection.[24] The degree of leachate penetration into the ground also exceeded the standard. Moreover, the Korea National Defense University and a large military oil tank, which was at risk of a potential explosion from the landfill's methane gas emissions, were located nearby. Despite these inappropriate conditions, the City appointed Nanjido as the Municipal Solid Waste Landfill to resolve the pressing waste management issue. Consequently, when Nanjido Landfill opened in 1978, military-related facilities surrounded it, and floods continued to damage the Landfill residence even after the 1990s (Figure 1.9).

Military facilities often need large-scale land for training as they create noise pollution with the blasts from shooting or aircraft training. Similar to military-related facilities usually located in or close to the city for the city's defence purposes, the city government has tended to build anti-social facilities like orphanages (for the abandoned population) and waste management facilities (for abandoned materials) within the boundary of government control so that they can serve city lives while simultaneously preventing such disagreeable entities from blending into other parts of the city. That is, the nation/city appropriates and colonises a portion of its own land along with the people who inhabit that territory to build indispensable yet abhorred facilities for the sake of the cleanliness, orderliness and wellbeing of urban life.

Environmental and social ecologies[25] of Nanjido Landfill

• Environmental ecology of Nanjido Landfill

As for the environmental ecology, the foremost dramatic changes of Nanjido during its active landfill period are its topographical transformations from wetland to levelled ground to high mounds within a 15-year period, and its geological change from earth to a mixture of wasted consumer goods that have caused earth, water and air pollution. During the first landfill stage (1978–1984), the City dumped garbage in the wetland and small stream of

Figure 1.9 Satellite photograph of Nanjido Landfill. The military-related facilities (concealed for national defence purposes) are located close to the landfill site. 1985. Map © National Map Museum

the Nanjido area until they were full in the mid-1980s. The city government initially planned to close the landfill in 1985 when refuse had completely filled the wetland, but it prolonged the landfill's use and commenced the development of an overground landfill because it had failed to prepare an alternative waste management policy by then. During the second phase of Nanjido Landfill (1985–1992), the garbage was piled so high that it eventually created two gigantic mounds that reached sea level at 94–98 m (80–84 m from the ground) in height and 2,720,661 m² in mass.[26] On the map of the mid-1980s, the stream and wetland filled up, and the maps of the 1990s and the 2000s show two overground mounds, marked as mountainous hilly sections[27] (Figure 1.10).

From the architectural landscape point of view, the landfill's identity—whether it is a built structure or a natural environment (as biodegradable waste assimilates into the existing nature over time), or both—is an essential subject of interest. A landfill is generally regarded as an artificial invasion of nature primarily because conflicting components of various types of waste meld into the earth. It is even more so in unsanitary landfills, where mixtures of waste are dumped together without any waste treatment systems.

Figure 1.10 Topographical changes of Nanjido. Four maps on top © National Map Museum; the first photo on the bottom left © Jong-Chul Won; second and third photos on the bottom © Seoul City; and the fourth photo on the bottom right © Jeong Hye Kim

During garbage decomposition processes, the earth experiences incessant bio-chemical turmoil. As for water pollution, the mixed garbage dump generates polluted leaking leachate in unsanitary landfills like Nanjido Landfill. The untreated leachate leaked into the Han River and severely polluted the environs, particularly the small stream of the Han River north of the Nanjido Landfill site.[28] According to a 1993 report published immediately after the landfill's closure, the leachate from Nanjido Landfill included poisonous heavy-metal elements, such as CN and lead, 20 times over the safety standard.[29] Also, the air pollution of mostly toxic gas caused concern not only for the damage it brings to the stability of the environment, but also for its generation of hazardous accidents (e.g. fires caused by methane gas leaking) that threaten people's lives.[30]

Although we tend to differentiate biodegradable garbage that decomposes naturally from non-biodegradables or pollutants, we can hardly distinguish one from the other in practice because we are neither fully aware nor knowledgeable of decomposition processes. The garbage types dumped in Nanjido Landfill were so diverse, ranging from construction waste,[31] used coal briquettes and food to everyday objects made of paper, plastic, fabric and metal. Due to the garbage collectors' task of sorting out recyclable non-biodegradable items from mixed garbage, a considerable portion of Nanjido Landfill became composed of biodegradable waste that was expected to merge with the natural elements in the long term.

Meanwhile, archaeological studies and chemical research suggest another opinion, asserting that biodegradation is not as active as we generally imagine. According to the archaeologist Rodolfo Lanciani's report (1890) on his excavation of an ancient Roman garbage dump on the Esquiline Hill, much of the garbage from imperial times had yet to decompose fully.[32] Based on these inexplicable archaeological discoveries' incompatibility with chemical theories, we cannot simply determine whether the landfill is an artificial intrusion into nature, or a conglomeration, or co-existence, of the artificial with the natural in conflict. The complex relationship between the landfill and the existing nature provides the foundation for a crucial reconsideration of the built and natural environments and their conflicting or complementary relationships with each other.[33]

- Social ecology of Nanjido Landfill

Nanjido Landfill was an inhabited landfill, where a substantial population of people lived and worked; over 4,000 people established the garbage collectors' community. Considering the socio-ecological facets of Nanjido Landfill will provide a lens through which we can examine the landfill within the social urban fabric of Seoul. We can consider the social ecology of Nanjido Landfill in terms of three aspects: 1) the settlement's legitimacy, 2) its habitability and 3) its social value. Regarding the settlement's legitimacy, despite their illegal residence in city-owned land during the early landfill period, the inhabitants were not merely squatting partly because they were involved in the municipal landfill's creation (e.g. top-soiling and hardening of the earth, etc.) and were responsible for the labour of recycling in the landfill. Regarding the habitability of the Nanjido Landfill region, it is hard to construe it as habitable or uninhabitable; compared to the so-called appropriately designed 'formal' urban space, the living conditions of the landfill area were uninhabitable and precarious because they lacked infrastructures such as water supply and sewage systems. However, seen from the landfill community's capacity for self-help, we cannot judge habitability only by the norms of the urban space of Seoul. Due to the strength of community within the landfill families—they had higher sociability in part because most of them shared common concerns regarding their precarious living and working conditions—it is also hard to conclude that the landfill residence was uninhabitable. In Nanjido Landfill, the legitimacy of settlement, habitability and social value are all related to one another, and they lie somewhere on the border between the legitimate and illegitimate, habitable and inhabitable and socially valuable and valueless. This uncertainty, accordingly, leads us to question if socio-ecological equilibrium actually corresponds to the degree of industrialisation and/or modernisation, which is generally calibrated by physical norms determined by the society.

In addition, we can also relate the socio-ecological aspect of the landfill, particularly its habitability and sociability, to the question of whether the landfill is a place or a facility. A landfill, in general, is regarded as a facility for waste management. Inhabited landfills, however, function as both facilities and places: places of residence and employment for garbage collectors, where many social activities take place amidst the facilities for waste management. The primary reason why Nanjido Landfill is not often discussed from a socio-ecological point of view is that it is only regarded as a facility, not as a human habitat. Similar logic is applied to the post-landfill period; on the one hand, after Nanjido Landfill was turned into Nanjido Post-Landfill Park, the social value of the community was replaced with the on-and-off social functions of the park visitors. On the other hand, the park is still a facility that requires the safe management of the closed landfill underneath it, which mainly meant maintaining the risk of explosion associated with the lingering methane gas. That is, as long as the landfill exists, the site has the dual identities of place and facility, which posits Nanjido Landfill as well as Nanjido Post-Landfill Park in an ambiguous position on the border. It also opens the way to viewing waste as an entity in a state of ambiguity.

Post-landfill period (1993–present): regeneration

Post-landfill vision in the neoliberal economic turn

At the turn of the twenty-first century, the re-illumination and regeneration of obsolete facilities of the industrial age for different use began in the new era. The overflowing unsanitary landfills in the nations of the West and some developing countries, amongst other facilities, were causes of major concern for urban planners and landscape architects because they occupy a large scale of land and their pollution deteriorates the urban environment. With advanced technologies, landscape architects devised methods to transform polluted landfills into landscape parks; for example, the transformation of New York City's Fresh Kills Landfill (1948–1997) into Freshkills Park (2010–present) is one of the exemplary cases.[34] We can also view Nanjido Landfill's transformation into Nanjido Post-Landfill Park in this globalised socio-economic context.

Following the outbreak of South Korea's IMF (International Monetary Fund) financial crisis in 1997/1998, Nanjido Landfill site's transformation into Nanjido Post-Landfill Park developed a complicated relationship to the shifting economic circumstances. At the end of 1997, the South Korean economy suffered from a severe shortage of foreign exchange on the brink of the national default. When the IMF bailed it out, the organisation required the South Korean government to implement economic restructuring plans, which resulted in two systematic changes: first, a turn to the ICT (Information and Communications Technology) industry and, second, a rapid metamorphosis into the neoliberal capitalist system. The new economic environment also entailed new urban development projects, aiming to draw foreign investment through the city's brand value. The construction of the Digital Media City (hereinafter DMC, 2002–present), a part of the Sangam-dong New Millennium Town development, was one of the representative cases. Nanjido Landfill's metamorphosis into a park is positioned in this overall transition from post/industrial capitalism to the neoliberal capitalism, which converts anything—from real estate to abstract concepts like 'green' or 'environmental awareness' and 'art'—into exchange value. Likewise, throughout the 1990s, with the delay of Nanjido Landfill's closure, the post-landfill plan took shape along the shifting economic circumstance that was rapidly turning toward neoliberalism. Besides the excessive amount of waste that triggered the landfill's closure, the demise of the landfill itself also led to the downturn of the local waste and recycling business. While the local recycling industry, operated by the garbage collectors' manual work, reached its peak in the early and mid-1980s, South Korea started to import lower-priced recyclable waste and the local recycling industry began to decline as the global market system dominated from the late 1980s. In this sense, the closure of Nanjido Landfill is partly a result of the globalised recycling market.

In 1989, Seoul City decided to close Nanjido Landfill within 3 years.[35] Although unofficial waste dumping continued until the mid-1990s, the City officially closed Nanjido Landfill at the end of 1992 and began its plans for the landfill's stabilisation[36] and regeneration. Due to the delayed relocation of the landfill inhabitants, however, the stabilisation and reclamation of the landfill actually began in 1997. As the Landfill's closure grew near, Seoul City published *The Preliminary Report on the Long-term Land Use of Nanjido Landfill*, which focused on the long-term development of Sangam-dong into a 'teleport town'[37] as in the cases of New York (Staten Island), London (The Royal Docks), Tokyo and Yokohama (Minato Mirai 21). This idea later became the basis for the development of the Sangam-dong DMC.[38] As such, Nanjido Landfill's transformation into Nanjido Post-Landfill Park must be viewed within the context of the overall Sangam-dong development plan and building of the 2002 FIFA World Cup main stadium in the area (Table 1.3).

At first, Nanjido Landfill's closure and stabilisation made it clear that Sangam-dong was the last remaining undeveloped town of Seoul, and added fuel to the City's decision to construct the World Cup main stadium in the area, which the City believed would stimulate regional development. When Sangam-dong was appointed as the 2002 FIFA World Cup main stadium site in May 1998, the City made an urgent decision to reclaim the closed landfill site as a landscape park. It then simultaneously proceeded with the landfill's stabilisation and reclamation and Nanjido Post-Landfill Park's landscape development to finish before the World Cup's opening in June 2002.[39] This, again, reinforced the Sangam-dong New Town Development Plan (1999), which was followed by the Sangam-dong New Millennium Town Plan of the next year (2000). Lastly, it accelerated the original Sangam-dong Housing Development plan, which included the building of the DMC in

Table 1.3 National events and their influences on Nanjido Landfill's closure, transformation into a park and overall Sangam-dong redevelopment

	Nanjido Landfill site development	*Sangam-dong area development*
1988	**1988 Seoul Olympic Games**	
1989	Decision to close Nanjido Landfill	
1991–1992	Landfill stabilisation plan (preliminary)	
1993	Landfill closed	
1993–1995	Stabilisation plan & residents' relocation	
1995–1996	Stabilisation designed & planned	
1996	**2002 FIFA World Cup Korea/Japan decided**	
1997	Stabilisation construction began	Appointed Sangam-dong as Housing Development Region
1998		Decided to build 2002 FIFA World Cup games main stadium in Sangam-dong area
1999	World Cup Park plan	Sangam New Town Development Plan
2000	World Cup Park construction began	Sangam-dong New Millennium Town Plan
2002	World Cup Park opened	DMC (Digital Media City) development began
2020	Complete landfill stabilisation	(Continue)

Table by Jeong Hye Kim

North Sangam-dong. While the building of the World Cup main stadium influenced Nanjido Landfill's transformation into a park, the park-building consequently increased both the real-estate and exchange value of the stadium site and the overall Sangam-dong New Millennium Town Plan. In short, each development not only had a mutual effect on the values of other projects, but also demonstrated its own symbolic and exchange value.

Aside from the direct environmental benefits for its regeneration of the degenerated environment, amelioration of environmental conditions and revival of nature, Nanjido's metamorphosis from a garbage dump to a revitalised landscape park produced added value to the site and surrounding regions. It is notable here that the natural environment has become an exchange value that is particularly tradable amongst the middle and upper middle classes. This is the point at which matters of environmental ecology fuse with those of social ecology. The unequal economic distribution amongst inhabitants of North Sangam-dong and Nanjido Landfill plays a role in the social ecology of the town development. The Sangam-dong development intended to resolve the problem of a housing shortage in Seoul by building an apartment complex on the site of the existing single houses in North Sangam-dong. The principle of this plan was to resettle current residents in the newly renovated environment of their current site of residence.[40] Seoul City's initial attempt to apply this principle to the Nanjido Landfill population as well as to the North Sangam-dong residents is notable. The government planned to demolish the existing illegitimate collective housing complex of the Nanjido Landfill residents and transform the site into a green zone, while building permanent rental apartments in the New Millennium Town to enable the landfill residents to stay in the region. In reality, however, the City did not build public housing and most Nanjido habitants could not afford to live in the new Sangam-dong apartments that had been built for the middle or upper middle classes. As a result, they were relocated to permanent rental apartments in different towns, while the existing Nanjido settlements were demolished and turned into part of Nanjido Post-Landfill Park.[41]

This demonstrates a recurring pattern of gentrification in global urban redevelopment. That is, the material value of the built environment and accordingly the real-estate price are raised, thereby replacing the impoverished community with that of the middle and upper middle classes, further marginalising the lower income class. For Nanjido Landfill garbage collectors, the closure of the unsanitary landfill and the community's disintegration into different regions meant unemployment and the loss of their foundation for a stable socio-economic life, a position that Zygmunt Bauman calls 'the wasted lives'. In addition, the Nanjido Landfill inhabitants' relocation to other regions has rarely been a subject of concern amongst civic/environmental organisations. They have not made this an issue mainly because Seoul City compensated the Nanjido residents with a relatively reasonable monetary amount for their forced eviction; therefore, most organisations regarded their situation as exceptional to that of other redevelopment cases, in which existing inhabitants would receive little or insufficient compensation. However, veiling the socio-ecological disintegration with financial compensation—often under the name of environmentally conscious spatial regeneration—could be more damaging in the long term than conflict over compensation because, once a deal is settled, there is hardly any willingness to rehash other fundamental problems.

Nanjido Post-Landfill Park[42]: cultural and environmental value creation

Nanjido Post-Landfill Park (opened on 1 May 2002) consists of five parks: Noeul Park (Landfill 1), Haneul Park (Landfill 2), Peace Park (Landfill 3), Nanjicheon Park (Nanjido Landfill garbage collectors' collective housing complex site) and Nanji Han River Park. Of these five parks, Noeul Park and Haneul Park cover the former landfill mounds where land stabilisation and regeneration are in progress with expected completion in 2020. We can view the regeneration of the unsanitary landfill site from two perspectives: material and immaterial cleansing. First, the material cleansing is the process of land rehabilitation by way of physical and bio-chemical engineering construction methods. Second, the immaterial cleansing not only removes the dirty and potentially threatening image of the landfill, but also develops the new value of the site. Seoul City expected cultural and environmentally conscious practices to reconstitute the site's image and produce heightened brand value for the city and the region, especially in the neoliberal economic context where everything becomes exchange value.

Since its opening in 2002, Nanjido Post-Landfill Park has been known for its cultured opportunities for leisure and environmental awareness; the SeMA Nanji Residency and the sites for leisure activities (e.g. campsites and playgrounds) represent the former and the Resource Recovery Plant the latter. While these facilities symbolically represent the city's endeavours to build cultural and environmentally conscious brand value, waste management has made the so-called useless or valueless entities invisible in the neoliberal era.

Located at the foot of Noeul Park, the SeMA Nanji Residency (2006–present) is the artists' residency programme facility run by the Seoul Museum of Art. Since 2007, the museum has been providing artists with studios and holding exhibitions and art events through the artist residency programme. As the leachate facility buildings were left unused thanks to other treatment systems, the new Seoul mayor Lee Myung-bak (2002–2006) and the then director of the Seoul Museum of Art Ha Chong-Hyun (2001–2006) agreed to renovate these unused structures for a museum-operated artist residency. By adding cultural value to the city government, it consequently turned out to be a successful project for the city's political economy. Apart from its well-managed curatorial programmes, the SeMA Nanji Residency as an architectural entity has had safety-related environmental issues that have not been clearly identified let alone resolved.[43] One of the senior researchers at the Seoul Institute, who co-researched *Evaluation of Nanjido Landfill and Environmentally-friendly Restoring Strategies* (2000), stated that the overall environmental conditions of Nanjido Post-Landfill Park meet the safety standards (not hazardous to human health), yet, it is still not recommendable to reside in the park because safety standards are relative to the socio-cultural context and every individual may react differently depending on his/her physical conditions.[44] This suggests the difficulty in determining absolute safety standards, which was particularly true of the early stages of stabilisation after the landfill closed. Although the Post-Landfill Park's environmental conditions satisfied the scientific standards for safety for 1 year of residency, the issue here is that the resident artists' sensory experiences of unidentified causes (e.g. gas or odour) are still tacit. The unexplored inconveniences are also related to the SeMA Nanji Residency's position since it was established in the context of a culturally and environmentally minded political economy; when culture and politics are interconnected, the exhibitionist political economy is apt to appropriate the former, which is likely to happen in a more subtle way within the neoliberal economy than ever before. Aside from the Nanji Residency's contribution to the art world, the institution's socio-political position represents how the neoliberal political economy co-opts cultural practice.

The Mapo Resource Recovery Plant (2005–present), located close to the SeMA Nanji Residency in Noeul Park, is a corporatised waste treatment and energy reclamation system. It is an advanced incinerator equipped with energy regenerating and resource reclaiming systems. Of the total 600–700 tonnes of waste treated per day, it incinerates 97% and landfills the remaining 3%. Since the plant's opening, the rate of emission of pollutants has not risen above the legal standard. It burns the garbage, collects the thermal energy generated during the burning process along with the methane gas extracted from the landfill and uses them as the electricity source for the North Sangam-dong residential and commercial area (DMC) (Figure 1.11).

However, there was a long argument between Seoul City and the indigenous North Sangam-dong residents surrounding the building of the Mapo Resource Recovery Plant for the decade from 1995 until 2005. For the residents, the plant was like an incinerator in that it could contaminate their living environment and debase the town's image, which would accordingly lower the brand value and real-estate value of the region. From 2004, the

Figure 1.11 Mapo Resource Recovery Plant. 2014. © Jeong Hye Kim

Sangam-dong Residents Support Committee inspected and monitored the plant's operation, such as the amount of waste treated per day and treatment methods, and suggested that the plant hire some residents to work at the company. The two partially negotiated these terms, and the agreement moderately worked.[45] Likewise, the Mapo Resource Recovery Plant has functioned well, practically and symbolically, as an environmentally conscious system for waste treatment in the new era. On a practical level, it has been operating without significant problems of pollution; on the symbolic level, it represents how the environmental industry can recreate the value of a site. Meanwhile, although the Nanjido Landfill residents desperately needed new jobs since the landfill's closure in 1992, the plant and committee had excluded them from the negotiation table. This exclusion, again, raises the question of whether or not the Nanjido Landfill inhabitants had a right to their environmental and social ecologies as citizen-subjects.

Considering the plant in respect of the waste treatment industry, it entailed the transition from an unsanitary landfill that relied on the manual recycling of self-employed garbage collectors to a corporatised sanitary waste management system.[46] Regarding this industrial transition, Saskia Sassen argues that the elite managers of multi-national corporations appear to control economic globalisation but the whole system will collapse without the support of traditional occupations involving physical and manual labour, which have been expulsed or currently rendered invisible.[47] The same holds true for this situation; manual recycling tasks, which garbage collectors mainly in Nanjido Landfill had undertaken, have now been allocated to invisible individuals (e.g. temporary workers hired by recycling corporations). Yet in public spaces, self-employed temporary garbage collectors still continue manual garbage collecting in informal ways.[48] As Saskia Sassen points out, this distribution of labour implies that a corporatised system does not completely replace manual tasks; rather, it only degrades the workers' socio-economic status and makes their work insignificant.

The Nanji Golf Course debate: ecology and class issues

In the landscape plan of Nanjido Post-Landfill Park, the top of the Noeul Park mound (Landfill 1) was initially designed as the nine-hole Nanji Golf Course. However, it is currently used as a landscape park with leisure facilities, including a campsite. Throughout the first round of decisions, the construction and the opening of the golf course along with its subsequent re-opening as a park for leisure activities, there had been a legal battle between Seoul City and the Korea Sports Promotion Foundation over the management of the golf course. Further complicating the situation, conflict over building the golf course in the park also erupted between the City and civic/environmental organisations. The golf course incident was remarkably the only case throughout the entirety of the Nanjido Post-Landfill Park development project with which civic/environmental organisations took issue.

Thus, the debate surrounding the building of the golf course and its closure demonstrates how civic/environmental organisations viewed Nanjido
Landfill's transformation into a park, interpreted the environmental and
social ecologies and identified the public.

Seoul City first announced its plan for the Nanji Golf Course in early
2000 and spent 17,000,000,000 KRW (17,000,000 USD) on its construction with the investment of the Korea Sports Promotion Foundation.
According to the landscape engineers, a golf course is technically appropriate for the closed landfill site during its stabilisation processes, and several countries like Australia had adopted this method.[49] Seoul City intended
to open the field to the public at a low-cost green fee so that they could
overcome preconceptions of golf as a luxury sport. In 2001, the City and
the Korea Sports Promotion Foundation made an agreement to co-operate
the Nanji Golf Course, and construction began after the park's opening in
2002. From 2004 until 2008, however, the two organisations were involved
in litigation over the green fee and ownership of field operations. In 2004,
the City enacted a green fee-related ordinance (15,000 KRW [15.00 USD])
just before the completion of the golf course and claimed the City's ownership based on the argument that the golf course was a 'public' facility. Upon
the City's decision, the foundation insisted that they must double the green
fee just to operate the facility realistically, and then, filed a lawsuit against
the City to nullify the ordinance. After years of battle in court, in 2008, the
two organisations agreed to close the golf course on the condition that the
City compensate the foundation for its investment funds in exchange for
complete ownership of the site. Several months later, the City finally re-
opened the field as the Family Park for picnicking and camping.[50]

As plans for the golf course proceeded, the Green Seoul Citizen
Committee[51] had argued against the idea and protested actively around the
time of its completion in 2004. When Seoul City first announced the plan
in 2000, 22 committee members objected to the decision and resigned from
the committee. In response to the protest, the then mayor, Go Kun, stated
that the City aimed to build the public golf course and maintain its field
in an environmentally responsible way by using organic pesticides to protect the soil from chemical pollution.[52] Upon completion of the golf course,
other civic/environmental organisations, including the Korean Federation
for Environmental Movements and Seoul Green Trust, issued the statement
that Seoul City must transform the golf course into a Family Park based on
the survey result that 87% of Seoul citizens agreed with them.[53] One year
later, the civic/environmental organisations held another press conference
to urge the City to change the use of the site. Their protest was grounded
on the idea that Seoul City must make the natural environment accessible
to more citizens, not limit its use to the upper class.[54] Overall, the Nanji
Golf Course had not become a significant issue—except during the Seoul
mayoral election campaign—until 2008 when the City nullified the golf
course plan. There is no doubt that the civic/environmental organisations'

protests affected the City's final decision to re-open the site as a public park. Nonetheless, we should reconsider the fact that civic/environmental organisations engaged with the Nanjido Post-Landfill Park's development only on the matter of the golf course construction, particularly when legal and political issues surfaced.

Throughout the conflict between Seoul City and the civic/environmental organisations, the public's right to the natural environment provoked the most concern. More specifically, the point of debate lay on the two sides' different understandings of the definition of 'public'.[55] The term 'public', which is often ambiguously confounded with 'mass', 'crowd' or 'citizen', is generally regarded as a group of people that seemingly includes 'everyone' based on the ideas of economic equality and political justice. In reality, however, the 'public' is used class specifically, almost always excluding classes under a definitive level of income. In the capitalist economy, for example, the 'public' connotes the population that is capable of properly producing and consuming, thereby, covertly excluding economically ineligible groups and rendering them invisible. In the Nanji Golf Course debate, while Seoul City asserted that it was a 'public' golf course open to all citizens who could afford the green fee of 15,000 KRW (15.00 USD), the civic/environmental organisations insisted that the site should serve the 'public', as only a limited number of people could truly access the golf course. From an ecological point of view, however, this debate hovers around the middle class's 'right to the green environment' and excludes the lower income class. This approach is ultimately rooted in a conception that regards the former Nanjido Landfill site only as a space for wasted materials, not as a human habitat, particularly one of the lower income class.

Environmental change always has an impact on the human habitat, more often on that of the socio-economically powerless. It is more so in developing countries where urban development projects take place so rapidly that they cannot conduct thorough environmental examinations, let alone fully apply the research results to the practice. This reality is essentially because they are initiated for political purposes; since most projects must be accomplished during a president or mayor's term of office, the administration exercises authoritative force, often damaging the natural environment and/or human habitats. In such cases, civic/environmental organisations and the media call attention to the environmental damage, but neglect the damage done to the human habitat especially when financial compensation is involved. In short, the politics of environmental regeneration, which is credited with the production of higher exchange value for the urban space, has become the primary matter of concern, eclipsing the importance of the embodied human habitat, particularly that of the lower income class.

The Nanjido Landfill residence's transformation into a park was at once a revival of the natural environment and a replacement of the human habitat with the exchange value of 'nature' for the 'public'—an economic class that could afford a clean and safe environment. Whether or not the transformation from a landfill to a park was positive is a question that requires examinations of the relational aspects of environmental and social ecologies.

Notes

1 Nanjido is geographically located downstream of the Han River in Seoul, South Korea. Topographically, it was a small island with wetlands along its shores. Administratively, the Nanjido area is 549 Sangam-dong, Mapo-gu ('dong' refers to village, and 'gu' means town) in West Seoul. Nanjido means an island of orchids (*nan*) and Lingzi mushroom (ganoderma lucidum) (*ji*). 'Nanji' means splendid beauty.

2 According to the US EPA, Municipal Solid Waste Landfill (MSWLF) refers to a discrete area of land that receives household waste. Household waste includes any solid waste, including garbage and septic tank waste from houses, apartments, hotels, motels, campsites and picnic grounds. See US EPA, 'Decision-Making Guide to Solid Waste Management', 1989; US EPA, 'Municipal Solid Waste Landfill Criteria, Subpart F—Closure and Post-Closure' (4 October 1993).

3 In this study, 'environmental friendliness', particularly in the post-landfill Nanjido Post-Landfill Park (World Cup Park) period, means minimising the artificial damage to the original state of nature and not polluting the earth, water or air of the human environment.

4 In this research, I use the term 'public' to express several different meanings: first, something related to government-led initiatives for the nation, city or large community's benefit; second, the large community as a whole; and third, civic, or something related to the citizen as a subject. In the third case, I use the term civic and public together selectively depending on the context.

5 The Ministry of Construction announcement No. 1031 (1963), No. 512 (1970) and No. 197 (1976) (Seoul City, *A Preliminary Report on the Long-term Land Use Plan for Nanjido Landfill*, 1992b: 51).

6 Nanjido Landfill's garbage workers, who moved into the area from the outside, are different from the indigenous residents of North Sangam-dong. I make this difference clear by calling the former the Nanjido Landfill inhabitants and the latter North Sangam-dong residents.

7 I call individuals who had lived in the region before the site turned into the municipal landfill 'indigenous Nanjido residents', distinguishing them from the Nanjido Landfill inhabitants who moved to Nanjido Landfill for garbage collecting.

8 Chan-gyun Shin, 'Nanjido, The Sound of Hoeing', *Gonggan*, No. 124, October 1977; and Il-cheon Shin, 'Democratic Republic Established by the Juvenile Vagabonds', *Sasanggye*, Vol. 10, No. 2, 1962.

9 The Samdong Institute, which had inherited the Samdong Boys Town, exists to this day under the new name of the Samdong Welfare Centre. Originally located in Nanjido, the Institute moved to the current Sangam-dong residential area, between the Digital Media City (DMC) and Nanjido Landfill Park.

10 In Lacanian psychoanalytic analysis, the 'imaginary' is partly illusory, leading to the misconception of oneself as a whole, but is necessary in establishing a meaningful relationship with the external world. Based on Lacanian psychoanalytic analysis and the neo-Marxian perspective of Cornelius Castoriadis, Ben Campkin adapted the notion of 'place imaginaries' to the urban context, which, he argues, can usefully articulate how we construct, recognise or distort contested sites (Ben Campkin, *Remaking London*, New York: I.B Tauris, 2013:

9–10; Jacques Lacan, *Écrits*, London: W.W. Norton, 2003 [1966] and Cornelius Castoriadis, *The Imaginary Institution of Society*, Cambridge: Polity Press, 1987 [1975]).

11 In the twenty-first century, the term 'green', unlike 'nature', is used with connotations of socio-political and economic values.

12 Opposing 'nature' with 'culture' is based on an anthropocentric and patriarchal viewpoint, against which scholars, including anthropologists and ecofeminists, have fought for decades during the late twentieth century. The oppositional categories of nature and culture arose as an ideological polemic in eighteenth-century Europe: a polemic, which created contradictions by defining women as natural (superior), but instruments of a society of men (subordinate). This contradiction persists into the twentieth century, setting up dichotomous metaphors in gender theory: nature vs. culture, wild vs. tame and female vs. male (Carol MacCormack and Marilyn Strathern eds., *Nature, Culture and Gender*, Cambridge: Cambridge University Press, 1980: 6–7). One branch of ecofeminism (e.g. deep ecology) focuses on the dichotomous split between nature and culture and how it enables the oppression of female and nonhuman entities. Ecological feminist Val Plumwood, in *Feminism and the Mastery of Nature*, states that the rationality of Western culture and elite domination have shaped the dualism between nature and culture, defining the former as the 'inferior' other (Val Plumwood, *Feminism and the Mastery of Nature*, London and New York: Routledge, 1993). For the implications of the woman-nature association in ecology, see also Kate Soper, *What is Nature?* (Oxford and Cambridge, MA: Blackwell Publishing, 1995: chapter 4). Soper explores the tensions between ecological 'naturalism' and the 'anti-naturalist' impulses of feminism and gay politics.

13 Soper, 1995: 195–196.

14 Regarding the matter of ecological equilibrium in agrarian life and a critique on the consumer capitalist system, see Agnès Varda's film *The Gleaners and I*, 2000.

15 The May 16 Coup in 1961 (led by the Major-General Park Chung-hee) led to the beginning of South Korea's dictatorial regimes, which led to the establishment of the Third Republic (1963–1972) and Fourth Republic (1972–1979) under the same President. After the Fourth Republic was terminated in 1979, President Chun Du-hwan took office and sustained ironhanded dictatorship until 1987 during the Fifth Republic (1981–1988), another military government.

16 www.nationmaster.com/graph/eco_gdp-economy-gdp&date=1962 and www.nationmaster.com/graph/eco_gdp-economy-gdp&date=1989 (updated in 2007, accessed on 1 June 2015).

17 www.nationmaster.com/graph/eco_gdp_percap-economy-gdp-per-capita&date=1962 and www.nationmaster.com/graph/eco_gdp_percap-economy-gdp-per-capita&date=1989 (updated in 2007, accessed on 1 June 2015).

18 Zygmunt Bauman comprehensively indicates material and social surplus, calling it redundant or the wasted. See Bauman, 2012 [2004], chapter 2.

19 Many apartments were built on top of these dumpsites. Most of the waste at the time, used coal briquettes and household trash, was often discovered when redeveloping the site. From the interview with Yong-soo Kim (16 May 2014), Seoul City public servant in charge of the Nanjido Landfill residents' relocation.

20 The Seoul Institute, *Evaluation of Nanjido Landfill and Environmentally-friendly Restoring Strategies*, 2000: 9–12. The Seoul Institute, formerly the Seoul Development Institute established in 1992 with approximately 250 researchers, is the research centre for Seoul's short-term and long-term urban development policies.

21 Seoul City, Gyeonggi-do province and Incheon City agreed to extend the use of Sudogwon Landfill (administratively located in Incheon City) until 2025. It was

originally supposed to close at the end of 2016. Before 2025, three local governments must set up alternative landfills in each region. See *Yonhap News* (28 June 2015) and *Korea Herald* (2 July 2015).

22 Regarding the diverse plans for Nanjido's use (e.g. building a stadium), see *Kyunghyang Shinmun* (3 August 1977) and *MK Business News* (7 February 1978).

23 Michael Pugh and Philip Rushbrook, *Solid Waste Landfills in Middle- and Lower-Income Countries*, World Bank Publications, 1999: 13.

24 The Ministry of Construction, South Korea, 1967.

25 Félix Guattari, in his discourse on three ecologies, mentions the media environment as a crucial register that affects human subjectivity and collectivity as an aggregated mass. In this study, social ecology focuses more on local politics and economic situations, yet it still argues to revive sociability and human subjectivity based on singularity, sharing the essential idea of Guattari's social ecology. See Guattari, 2000 [1989].

26 It is one quarter of the size of Fresh Kills Landfill in New York. Including the slope areas of the Nanjido Landfill mounds, however, Fresh Kills Landfill is approximately 2.5 times the size of Nanjido Landfill. Considering that Fresh Kills Landfill was in operation for 54 years (1947–2001), the mass of garbage in Nanjido Landfill is equivalent to that of Fresh Kills Landfill.

27 Nanjido Landfill's garbage collectors called the landfill mounds the 'garbage mountains'. See Byung-cheon Choi, 'Nanjido Report', *New Family*, Vol. 369, May 1987: 34–42; Sun-hee Lee, 'Seoul Landfill, Nanjido', *Saemteo*, Vol. 19, No. 12, December 1988: 10–14; and Un Song, 'The Last Winter of Nanjido Children', *Our Mid-level Education* 24, February 1992: 112–117.

28 Interview with Nun Magdalena (26 August 2014) who lived and worked as a garbage collector in Nanjido Landfill for 3–4 years in the late 1980s.

29 *MK Business News* (7 May 1993).

30 Once a fire broke out in Nanjido Landfill, which was full of inflammable materials and gases, they had to cut off the whole mass of the landfill to block the fire from catching on another mass of land. From the interview with Yong-soo Kim (16 May 2014).

31 On 29 June 1995, the Sampoong Department Store collapsed due to constructional problems. Of the total 502 dead, most of the corpses were buried in Nanjido Landfill (Seoul Foundation for Arts and Culture ed., *1995 Seoul, Sampoong*, 2016: 179).

32 William Rathje, Cullen Murphy, *Rubbish! The Archaeology of Garbage*, New York: Harper Collins Publishers, 1992: 112–113.

33 Richard Ingersoll discusses this in 'The Ecology Question and Architecture' in Greig Crysler, Stephen Cairns and Hilde Heynen eds., *The SAGE Handbook of Architectural Theory*, London: SAGE Publications, 2013 [2012]: 573–589.

34 Mohsen Mostafavi and Ciro Najle eds., *Landscape Urbanism*, London: Architectural Association, 2003; Charles Waldheim ed., *The Landscape Urbanism Reader*, New York: Princeton Architectural Press, 2006; Caroline Klein et al. eds., *Regenerative Infrastructure*, Munich, London and New York: Prestel, 2013; and James Corner, *The Landscape Imagination*, Alison Bick Hirsch ed., New York: Princeton Architectural Press, 2014.

35 In 1989, after the Seoul 1988 Summer Olympics, Seoul City considered the citizens' complaints about the distasteful smell from the overflowing amount of garbage of Nanjido Landfill, and then, decided to close the Landfill within 3 years.

36 Stabilisation means turning the polluted earth, currently in active chemical processes, into a secure state. Filtering and purifying leachate and collecting methane gas are two major processes used to stabilise contaminated landfills. See Ernest

Lehmann ed., *Landfill Research Focus*, New York: Nova Science Publishers, 2007.

37 For related studies, see Sun-joo Park, 'Nanjido Teleport Development Plan', Master's thesis at Seoul National University, Graduate School of Environmental Studies, 1993; and Seoul City, *A Study on the High-density Business Town Development of Seoul*, Seoul City, 1992c.

38 Seoul City, 1992b: 23–78.

39 Considering that the Park landscape design was changed several times, the actual construction endured for only approximately 1 year. From interviews with Gye-dong Ahn (11 August 2014) who managed the Nanjido Landfill Park landscape design and with Prof In-sung Lee (27 August 2014) who supervised the Park construction.

40 An apartment building holds many more households than a house on a site of a similar scale. In the late 1970s, the average ratio of the floor area to the building site of an apartment was 163%, whereas that of a detached house was 50% (Seoul City, *A Preliminary Urban Planning of Seoul City*, 1990: 79).

41 Ibid.: 47, 66, 102–109.

42 During the initial stage of park planning, Seoul City named it 'The Millennium Park' but they changed it to the 'World Cup Park' before the Park's opening.

43 Because of the sensitive issue of 'residing' in the closed landfill site for a month or a year, unlike the English name, SeMA Nanji Residency, the Korean name is Nanji Art Creation Studio.

44 Interview with Dr Woon-soo Kim (22 April 2014), senior researcher of the Seoul Institute.

45 Interview with Deok-Hee Kang (28 April 2014), manager of Korea Federation for Environmental Movement.

46 Seoul City has commissioned the Plant to a private waste management company via larger agency corporations like GS Construction, Hyundai Construction or Hyundai MOBIS.

47 See Saskia Sassen, *Expulsions*, Cambridge, MA: Harvard University Press, 2014; Saskia Sassen, *The Global City: New York, London, Tokyo*, NJ: Princeton University Press, 2001 [1991]; and Hong-Bin Kang, 'Planned Development of the "Creative Milieu" and its Sustainability', *Seoul City Research*, Vol. 11, No. 2, June 2010: 264.

48 On the history of garbage collecting of Korea, see Soo-jong Yoon, 'Ragpickers and Nation', *The Radical Review*, No. 56, Summer 2003b: 265–296.

49 See the Seoul Institute and University of Seoul, *Envisioning Millennium Park*, Proceedings of the International Symposium 'Towards the Sustainable Development of Nanjido', 1–3 December 1999.

50 *Yonhap News* (30 October 2008). During the years of court battle, Seoul City opened the golf course to the public for free (October 2005) and operated it for 2.5 years, enduring the deficit.

51 The Green Seoul Citizen Committee is a form of governance, consisting of about 100 members from different social groups, including experts from Seoul City, corporations and environmental and/or civic organisations. It was established in 1996 with the aim of promoting environmentally conscious sustainable development concerning Seoul City's urban policies.

52 *Dong-A Ilbo* (6 September 2000) and *Yonhap News* (9 May 2001).

53 *Hankyoreh* (9 June 2004) and *Dong-A Ilbo* (13 September 2004). The survey was conducted by the Citizens' Institute for Environmental Studies.

54 *Kyunghyang Shinmun* (9 June 2005).

55 Regarding the relationship between the City and the Committee, see *Yonhap News* (31 August 2001).

2 Sanitary management in post-war Seoul

Chapter 2 explains the meaning of sanitation and patterns of sanitary management in the 1960s and 1970s before Nanjido Landfill's opening in 1978. In each part, I examine the political and social norms that determined the meaning of waste or dirt in the post-war urban space of Seoul and show how government-led waste control practices during the 1960s and early 1970s—for example, fumigation with chemicals—influenced the operation and maintenance of Nanjido Landfill. As society perceives material and social waste as analogous, I deal with the control of both forms of waste in this study of sanitary management.

First, I question the meaning of cleanliness or sanitation in South Korea during the post-war era: sanitation as the foundation for the physical health of human resources essential to industrial and military power, and for the mental health associated with anti-communist ideology. Second, I examine how South Korea, particularly Seoul City, oversaw ragpickers[1] as a method of sanitary management in the 1960s and 1970s. This will demonstrate that the national/municipal government sustained the sanitary urban space by cleansing material and social waste simultaneously. The historical lineage between the early ragpickers and the Nanjido Landfill garbage collectors is significant because a proportion of the ragpickers later constituted the initial population of the Nanjido Landfill community. Third, I scrutinise the active use of DDT as another method of sanitary management in Seoul, focusing on fumigations with the insecticide for the 20-some years since the Korean War, while discussing it in close connection to the global context of environmental concern, which was historically marked by the publication of Rachel Carson's *Silent Spring* (1962).[2] Fourth, based on these sanitary practices and the public's fears and beliefs about fumes (both anxieties about and reliance on chemical fumes), I attempt to define the identity of Nanjido Landfill: a borderline character, rooted in the conflicting fumes of the landfill site.

Drawing from anthropologist Mary Douglas's ideas of 'purity' and 'dirt' demonstrated in her book *Purity and Danger* (1966),[3] Zygmunt Bauman developed the notion of 'dirt' in the globalised socio-economic context, cautioning that there are no absolute criteria for purity or patterns for maintaining a pure state. He explains:

> From Mary Douglas's analysis, the interest in purity and the obsession with the struggle against dirt emerge as universal characteristics of human beings: the models of purity, the patterns to be preserved change from one time to another, from one culture to another—but each time and each culture has certain models of purity and certain ideal patterns to be kept intact and unscathed against odds.[4]

Regarding post-war Seoul, political and socio-economic purity and the counter notion of dirt are related to the nation and city's reconstruction, which was based on anti-communist ideology and industrial development. As national security against communism, locomotive industrialisation and urbanisation had overlapped South Korea's compressed period of modernisation,[5] particularly in the 1960s and 1970s. We must consider the model of purity and patterns for its preservation in relation to the ideals and policies of the nation and city at that time. Therefore, examining how the national government of South Korea and municipal government of Seoul built a modern nation and city through sanitary management after the Korean War will enable us to figure out post-war Seoul's model of purity or cleanliness and the patterns the city intended to preserve. Examining the norms of purity or cleanliness and the methods of sanitary management will also help us to identify Nanjido Landfill as a site related to the cleansing of both the materially and socially wasted, within the modernising urban space of Seoul.

Sanitation as morality and ideology

Throughout the latter half of the twentieth century, South Korea went through turbulent political changes.[6] During the post-Korean War era, mostly governed by military dictatorial regimes, the national and municipal government endeavoured to set up social orders by clearing away anything perceived as a threat to establishing a modern nation and city. Cleaning, here, meant explicitly national or city-led social cleansing through physical (bodily and environmental) sterilisation. Amongst other practices, protection from infectious diseases was of particular importance on the national level because human bodies were major sources of labour for industrial development and anti-communist national defence in the military forces throughout post-war South Korea.

In this part, I look into modern Korea's use of the notion of sanitation as a method of moral and ideological discipline fostering healthy human resources. First, I investigate how the nation and Seoul City played a leading role in retaining human resources for industrial and military forces under the slogan of sanitation. Second, I explore how the government actively used the concept of sanitation for mental health in the era. That is, the anti-communist ideology of post-war South Korea, which was geo-politically positioned on the frontline of Northeast Asia, confronted the communist bloc of the Cold War era.

Sanitation as morality for industrial and military forces

Throughout the high industrial era of South Korea in the 1960s and 1970s, the national government's sanitary policies and campaigns focused on the battle against epidemic diseases to retain healthy labour resources. After the Korean War and throughout the 1960s, the national government publicly announced hygiene guidelines every summer to protect against the spread of encephalitis; guidelines included spraying DDT (along with lindane) on filthy sites in public spaces, such as toilets, garbage bins and sewers. The nation also published guidelines for daily domestic activities, particularly to prevent encephalitis infection.[7] Along with the national campaigns and instructions, the media often advised the public to preserve the clean and safe environment mainly by killing contagious insects such as flies, mosquitoes, fleas, lice, fall webworm and longwings, and by protecting themselves against diseases such as the flu, food poisoning, cholera, diphtheria and scabies. These national campaigns and instructions, promoted by the media, aimed to enable the government to manage hygienic issues in the private realm.[8]

Aside from the nation-led sanitary practices, each city facilitated its own hygienic policies. Throughout the 1970s, President Park Chung-hee emphasised that autonomous local governments must be responsible for the prevention of epidemics, such as that of cholera.[9] As a rapidly growing capital city in terms of both population and economy, Seoul, amongst other cities, was outstanding in its independent hygienic policies and practices. The City's most notable sanitary practice was the almost regular city-wide fumigations in public spaces.[10] Additionally, Seoul City announced its own epidemic prevention policies and issued emergency epidemic prevention ordinances as the nation had.[11] Its epidemic prevention department announced ten rules for individual healthcare: 1) take a shower after sweating profusely; 2) do not expose yourself to the sun for a long time or exhaust yourself; 3) remove weeds, making sure to clean them thoroughly, and water that has ponded; 4) do not get bitten by mosquitoes; 5) wash hands and feet when returning home; 6) kill flies and mosquitoes using insecticides; 7) avoid crowded places; 8) be aware of food safety; 9) be conscientious of children's healthcare; and 10) sterilise clothes by drying them in the sun[12] (Figure 2.1).

Seoul City's list of sanitary instructions was a guide not only for the creation of a clean and safe public domain but also for the cultivation of the healthy productive citizens essential to the modern industrial society. Regarding sanitation as a moral issue transferred the hygienic matter from the public sector to the private sector, which led citizens to internalise the ideas. Moreover, domestic behavioural instructions helped the national

Figure 2.1 Morning cleaning in Jung-gu (downtown), Seoul under the slogan of 'Spring Cleaning Week: New Town'. 26 March 1964 © Seoul Museum of History

government transfer the responsibility of public hygiene to individuals as a moral obligation. In other words, sanitation became a requisite for their contribution to the nation's industrial development as responsible modern citizens.

Historically, the relationship between cleanliness and the making of a modern nation dates back to the late nineteenth century. Confronting the imperialistic powers, the Korean Empire (1897–1910) initiated educational policies for the moral discipline of modern citizens by publishing *The Textbook on Moral Training* (*su-shin-gyo-gwa-seo*).[13] It was a moral guideline advising the citizens to nurture their individual healthy bodies, which were indispensable to the nation's industrial and military independency[14]; for example, the first chapter 'Body' explains how to train each body part, emphasising the importance of proper walking, sleeping, exercising, eating, clothing, bathing, residing and so forth:[15]

> Without a healthy body, you cannot achieve your individual goals nor can you contribute to the family's reputation. Dying young or becoming dejected is not only the renouncement of your own happiness but also the betrayal of loyalties to the nation and filial piety to your parents.[16]

Moral education prevailed throughout the Japanese colonial period but the purpose of the discipline shifted from fostering patriotic citizens who could protect the Korean Empire from imperialistic threats to raising colonial citizens, i.e. sub-Japanese compatriots who could facilitate the foundation of the Japanese Empire.[17] Additionally, the education of the Korean Empire is based on the moral lessons of Confucianism; during the colonial era, the imperial government systemised the moral education for hygiene and physical health based on centuries-old biological studies so that they could strategically apply it to the individual bodies of the colonised citizens. In the West, particularly in nineteenth-century France, people tried to distinguish moral education from society's preoccupation with hygiene (as insisted by David Émile Durkheim), but the connection between the two is hard to deny, at least in modern Korea.[18]

Although Korea's national policies for moral-hygienic education in the late nineteenth and early twentieth centuries were based on different philosophical bases and intended for different political purposes, the two policies were inextricably intertwined. The same was true of the national/municipal government's hygienic policies in the post-war Republic of Korea; in the new era of an independent South Korea, hygiene functioned as an individual's moral obligation and basis for each citizen's contribution to the nation/city's ideal—the industrialised and modernised nation or city. In essence, the policies implemented disciplinary education on hygiene as a moral responsibility on the premise that an individual's body is subordinate to the governing body, since it functions as a constituent of the socio-political system of the nation or city.

Sanitation as anti-communist ideology

Throughout its unique political situation during the post-war era, South Korea emphasised the importance of healthy bodies and minds rooted in anti-communist ideology. In other words, dirt, or the unhealthy body, was not only a threat to the building of a modernised nation, but also an element that could usher a political crisis into the nation. We can detect associations between notions of sanitation and anti-communist ideology throughout the 1960s and 1970s in: 1) the principles of national reconstruction announced by the major council of the administration and 2) the government policies that linked epidemic prevention to anti-communist ideology.

Immediately after the military coup in 1961, the head council established the headquarters of the Reconstruction Citizens Movement (16 May 1961–16 December 1963) and its sub-organisations in all towns across the country to proceed with diverse development movements, including educating citizens, developing rural areas and controlling vagrants under the slogan 'Accomplish the Welfare Nation'.[19] The socio-economic reconstruction movement was one of the measures the new governing military regime used to complete its coup.[20] Scrutinising the ideas of the Movement Office's three leaders sheds light on the spirit and purpose of the Movement; they mutually emphasised that 'individual autonomy that conforms to the regulations of society' is the primary virtue of modern human beings.[21] Furthermore, the objective of the Movement lay in increasing national power to accomplish the independent and autonomous unification of the divided country.[22] To that effect, the Office, then, announced seven practical guidelines: 1) counter the pro-communist ideology; 2) live humble and moderate lives; 3) strengthen the will to work; 4) encourage a spirit of production and construction; 5) cultivate the citizens' morality; 6) purify the mind and emotions; and 7) strengthen the citizens' body/health. These guidelines confirm that the nation managed the citizens' healthy bodies, honourable temperaments and emotions as requirements for the productive work force necessary to raise industrial productivity. Above all, it is notable that they set anti-communism as the primary guideline, which confirms the indoctrination of anti-communist ideology in the work ethics. Likewise, the anti-communist ideology of the Cold War era formed the bedrock of the Reconstruction Citizens Movement, which aimed to consolidate its doctrines.[23] In this sense, we can argue that 'dirt' in the post-war space of South Korea included communists along with the mad, the sick and the poor.

Furthermore, in the post-war space of South Korea, the government devoted all of its power and resources to preventing an epidemic, which could cause fatal damage to the nation's reconstruction and industrialisation. During this period, the government notably related epidemic prevention policies to anti-communist ideology, linking bodily threats to mental and political ones. In practice, President Park Chung-hee actively intervened in the nationwide epidemic prevention policies since his inauguration in 1963, and his intervention grew in significance after 1970, especially during his last regime (the Fourth Republic, 1972–1979).[24] In 1970, during his New Year's speech on administrative policies, the president mentioned healthcare and epidemic prevention as one of the three most important national policies, along with anti-communist national security and economic development.[25]

> The administration will try to preserve human labour resources by protecting citizens from diseases and reinforce the epidemic prevention systems to prevent the outbreak of tuberculosis and other infectious diseases.[26]

This demonstrates that disease control in the citizens' healthcare policies was designed to preserve human resources as the foundation for not only economic development but also national security against the communism of North Korea. Throughout the 1970s, overall hygienic policies were concentrated on epidemic prevention and practised by more systemised organisations. In preparation for the summer monsoon season, the national/municipal government set up systematic emergency task force teams to prevent a potential epidemic. For instance, in 1970, Seoul and its satellite cities organised the Emergency Epidemic (cholera) Prevention Association[27]; in 1972, the national government inaugurated the Special Task Force for Epidemic Prevention[28] and granted awards for their remarkable contribution to the prevention of epidemic diseases.[29] Additionally, in 1970, the national government and the Ministry of Health and Social Affairs of Korea proposed to grant the epidemic prevention agents immunity from military service:

> *In order to defend against North Korea's germ warfare attack, we have to build a strong prevention system.* To do so, we need to mobilise a sufficient number of epidemic prevention agents [...] for those service agents, we are considering granting immunity from military service.[30] (Italics added)

Although this agenda was not legitimised, it shows that the national government associated epidemic control with national security, creating connections amongst cleanliness, bodily health (military force), mental health (anti-communist ideology) and national security. The nation also established a relevant organisation as part of the military system, regarding the task of epidemic prevention as equivalent to that of military service.

These concentrated efforts to systemise the epidemic prevention organisations in the early 1970s transpired in response to the immense damage caused by the cholera outbreak of 1969 (from 17 August to 4 November 1969).[31] Since a Japanese sailor, the presumed first carrier, arrived in the port of Gunsan City (South-western part of Korea), an unidentified disease spread throughout the country. On 3 September 1969, the Ministry of Health and Social Affairs of Korea first announced that it was food poisoning caused by Vibrio infections, but it declared the disease was a new type of cholera on 9 September 1969.[32] This epidemic caused 1,528 casualties, including 137 deaths, amounting to a mortality rate of 8.8%.[33] This incident was recorded as the second most damaging cholera epidemic to have occurred since Korea's liberation. The military government seized the nationwide chaos caused by this epidemic as an opportunity to turn the citizens' attention toward external enemies and reinforce the anti-communist national security. Since 3 February 1970, when the media began to report that North Korea illegally imported pathogens from the Japanese corporation Yanagida, the national government insisted on a connection between North Korea's import of infectious agents and the 1969 cholera epidemic in South Korea. Then, it finally announced that North Korean spies had spread the cholera bacterium throughout South Korea.[34] In short, the military government used the cholera epidemic as a justification to reset the social system with defence purposes at its core.

Linking health issues to the political agenda is a method of fear politics. The governing power attempted to turn the citizens' attention from internal politics to the potential problems caused by external enemies—who have almost always been North Koreans—to perpetuate the long extended dictatorship, and to enlist the citizens' support of the military government and its stronger anti-communist policies. That is, in the post-war space of South Korea, especially during the latter phase of the dictatorial regimes, germs were regarded not from the perspective of health and wellbeing, but from a political and ideological angle. In summary, in the context of the post-war ideology, the healthy or unpolluted body and mind corresponded to anti-communism, while the unhealthy body and state of mind were deemed polluted and potential threats to the national security. The management of the citizens' physical and mental health and its link to the nation's economic growth and security effectively served to buttress the governing power of the military regimes.

Control of garbage collectors: physical sanitary management

The history of control over ragpickers began with the national/municipal governments' thorough control over the vagrant-homeless-ragpickers. The national government organised them as a professional part of the Reconstruction Workers Group (*geun-ro-jae-geon-dae*) immediately after the May 16 military coup in 1961 and called them 'garbage collectors' (*pye-poom-su-jip-in*). While control over garbage collectors transferred between the government and private organisations several times, management of the garbage collectors persisted until the mid-1990s. However, the opening of Nanjido Landfill was a decisive moment in their history, as it divided the garbage collectors into two groups: those who moved into Nanjido Landfill to live and work independently yet within a community of their own, and those who remained outside Nanjido Landfill either under institutional control or as independent waste workers.

This part reviews the work situations of ragpickers in the post-war space, and then, examines the two systems controlling the garbage collectors: first, the direct institutional control of the 1960s and 1970s; second, the indirect market control of the late 1970s onwards. This will elucidate how government and non-government organisations treated garbage collectors and looked upon them as wasted lives (social waste treatment), while equating them with their work, which dealt with wasted objects (material waste treatment).

Ragpickers in post-war Seoul

The history of ragpickers and their job of ragpicking in the public spaces of Seoul after the Korean War are closely related to the government's treatment of war orphans, who had largely become homeless. In Korean, homeless is called *roh-sook-ja*, which means a person who has no home, and thus, implies that he or she finds accommodation in the streets. Unlike *roh-sook-ja*, the homeless of the post-war era from the late 1950s until the 1970s were referred to as *bu-rang-ja (bu-rang-a)*, which literally means a vagrant who wanders around the streets with or without accommodations. In Seoul, most vagrant people at that time were war orphans and individuals who had fled their homes in rural areas to seek a new livelihood in the capital city of Seoul. Meanwhile, ragpicker (*nung-ma-ju-i*) in Korean pertains to a person who collects old clothes, wasted papers and other recyclable items in the streets (Figure 2.2). During the 1960s–1970s, ragpicking was considered one of the jobs of the lowest income class.[35] Although the term 'ragpicker' is not synonymous with 'homeless', the two are inseparable as many homeless people worked as ragpickers to achieve economic independence. The ragpickers' group was, in reality, almost indistinguishable from that of homeless people.

After the Korean War, approximately 120 groups of ragpickers existed in Seoul, and their number increased to 140 by 1961. Each group consisted of a leader (*jo-ma-ri*) and ten to 20 healthy teenagers of around 17 to 20 years old. The young ragpickers collected wasted papers from the early morning hours and sold them to the leader; the leaders then resold them to secondhand shoppers who finally resold the items to paper manufacturers.[36] The average daily collection of wasted papers gathered by a single group was about 15 trucks[37] (Figure 2.3).

The ragpicker groups congregated around Seoul and mainly worked downtown and in residential areas, where they could find more waste. Although ragpickers mostly picked up waste early in the morning to collect the previous day's refuse, their working hours were not necessarily limited or fixed; they could work at all hours, roaming the city streets. Exposed in public spaces while they worked, the ragpickers mingled with other members of the city in the post-war space of Seoul.

Institutional control of garbage collectors

Government control of ragpickers began in 1961 as a part of the Reconstruction Citizens Movement led by the military authority.[38] One of the major tasks of the Reconstruction Citizens Movement was to clear gangsters and vagrants, including ragpickers, shoeshine boys, homeless people and lepers, from public spaces by organising them under administrative control.[39] Although it was a national policy and each province was responsible for cleansing these populations, the practice was essentially concentrated in the capital city of Seoul.

One month after the coup (17 June 1961), the Ministry of Social Affairs of Korea announced that ragpickers must register to receive official permission for their jobs from the City (by 25 June 1961) and wear uniform with a nametag. Under this policy, the government officially banned the creation of private ragpickers associations and their group-based exploitation. Since the policy was announced, 882 people registered within a week and

Figure 2.2 A group of ragpickers (organised as a part of the Reconstruction Workers
Group) stationed in the Namdaemun region of downtown Seoul (1963).
© Seoul Shinmun

Figure 2.3 Ragpickers groups before institutional control (1953–1961). © Jeong Hye Kim

more followed. Then, in collaboration with the Ministry of Social Affairs, Seoul City set up tents as collective housing units outside downtown Seoul (Seongsu-dong [east], Jeongreung-dong [north], Sangdo-dong [south], Hongje-dong [west]) and designated the ragpickers work sites. On 1 July 1961, the City gathered the registered ragpickers for the first time and gave them the official title 'garbage collectors' (*pye-poom-su-jip-in*) (Figure 2.4).

The next year, on 22 February 1962, the municipal police department took complete control over the garbage collectors with the support of the national budget; they received 1,380 garbage collectors and accommodated them in 52 collective housing tents built in 11 districts of the Seoul City Police Department.[40] Finally, on 14 May 1962, the official inauguration of the Reconstruction Workers Group (garbage collectors amongst other professions) was held behind Seoul City Hall under the slogan 'Garbage Recycling'(*pye-poom-jae-saeng*), and 1,500 registered garbage collectors had participated[41] (Figure 2.5).

Initially 1,380, the official number of garbage collectors in Seoul increased enormously for the next 10 years; in 1975, the total number of city garbage collectors reached over 4,000 (the 3,510 registered and 500–600 unregistered combined).[42] Along with the police department's regulation of the garbage collectors' lives and work, religious organisations had been involved with the garbage collectors from the outset; the Christian

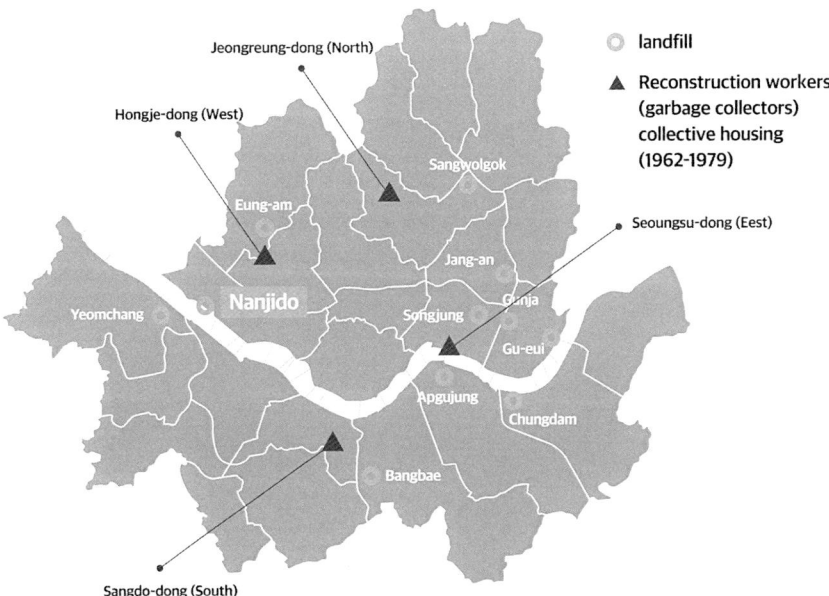

Figure 2.4 Control of garbage collectors: as a part of the Reconstruction Citizens Movement, the Ministry of Social Affairs of Korea in collaboration with Seoul City accommodated the registered workers (titled 'garbage collectors') in collective housing tents. 1961. © Jeong Hye Kim

Association was particularly in charge of the Reform and Discipline Council for Reconstruction Workers Group (*gueon-ro-jae-geon-dae gyo-hwa-ji-do-wi-won-hoe*) that strived for the garbage collectors' mental reformation.[43]

Likewise, the objectives and tasks of the Reconstruction Citizens Movement included both the physical and mental control of the subordinated workers. We can interpret the institutional control over the garbage collectors, more specifically, over their bodily and mental health, from Foucault's point of view, as the state's biopolitical organisation or population, mainly the birth and death of its population.[44] The physical control of the garbage collectors (along with the mad and the sick, e.g. lepers) demonstrates the national/municipal government's active use of biopolitics to produce imaginaries of an industrialised and modernised city without inappropriate social members (Figure 2.6).

Concerning the work processes of the garbage collectors (hereinafter garbage collection associations will be referred to as 'the Group' accordingly), each individual was supposed to collect 37.5 kg of refuse per day; since their large baskets could hold 19–23 kg of waste each, they had to fill two baskets daily. Specialised paper dealers (*eop-ja*) and other traders sorted the collected items and resold them to second-hand shoppers so that they could

Figure 2.5 Inauguration of the Reconstruction Workers Group (the team of garbage collectors) held on 14 May 1962 behind Seoul City Hall under the slogan 'Garbage Recycling'. © Seoul Museum of History

send them to paper manufacturing companies every 15 days.[45] However, the contract between the police department, second-hand shoppers and manufacturers resulted in a monopoly of the companies that often exploited the Group by paying them lower fees than that of the market price. Some dealers required premium fees from garbage collectors for allowing them to work in certain waste dumping sites or in large buildings.[46] To avoid the systemised corruption of the government and waste-related companies, many workers gradually left the Group and attempted to do their job independently outside the regions designated for garbage collection. Consequently, the government's control over the garbage collectors drove them outside the spatial and social boundaries of the urban space.[47]

Meanwhile, the disciplinary measures of private religious organisations supplemented the government's control by forcing the garbage collectors to carry out the more systematic sanitisation of public spaces. These methods meant the garbage collectors would not pollute the city environment with their presence, or visually expose themselves in public spaces. In other words, whether at the hands of the government or private organisations, control over garbage collectors reinforced the social imaginary of disgust toward them and their profession. For example, in 1976, garbage collecting

Figure 2.6 Following its policy to cleanse the urban space, Seoul City relocated the vagrants from the streets to a collective camp. 10 April 1964. © Seoul Museum of History

was banned downtown, an area that saw more visits from foreigners than other areas did; later, during the 1986 Asian Games and Seoul 1988 Olympic Games, garbage collectors were prohibited from going out in the streets.[48] Overall, as an impoverished groups, along with the mad, the sick and the communists, the garbage-collecting population needed to be cleansed from the public realm. In this sense, we can view the institutional control over the garbage workers as a demographic policy that consequently determined their identity: socio-economically inappropriate members (or non-members) of society, distinguished from the appropriate citizen-subjects.

In summary, in the modern industrial city of Seoul, maintaining the cleanliness of public space through sanitary management was one of the most substantial tasks of governance. The institutional control over the ragpickers demonstrates that the national/municipal governments actively exploited biopolitics to produce imaginaries of the industrialised and modernised city. The controlling authority's association of the socially wasted populations with wasted materials is also significant since their control simultaneously cleansed and rendered the garbage collectors invisible. The garbage collectors' responsibility for the material cleansing of the city space is symbolic as they are a socially marginalised group conceived as social waste. Therefore, this study of landfills and the garbage collectors in the landfills recognises both material and social waste concurrently, as the society had recognised and fused these two into one.

Market control of garbage collectors

During the institutional management of garbage collectors throughout the 1960s and 1970s, control was administered through spatial practices to a certain degree, but their residences and workspaces still merged with other spaces within the city. In the late 1970s, government-led public institutions gradually loosened their control over garbage collectors, and corporate market systems took over control of the urban planning and zoning, which gradually separated the garbage collectors' living and working sites from urban spaces for the supposedly appropriate members of the city.

In June 1979, the last year of the dictatorial regime of the Fourth Republic of Korea and a year after Nanjido Landfill's opening, the police department regained control over the garbage collectors, who had been in the hands of private organisations since 1975. The police department registered the garbage collectors in a new system under the name Voluntary Regeneration Workers Group (*ja-hwal-geun-ro-dae*)—using 'voluntary workers' was an effective strategy on the part of the controller as it could delude the workers with the misconception that there is no external or internal authority imposed upon them. 4,431 workers across the nation initially registered in the new Group. In Seoul, the City built pre-fab collective houses with shared kitchens and bathrooms to house 2,620 workers. However, this turned out to be an even more controlled camp system, in which the workers, especially low-income class working teenagers like ragpickers and shoeshine boys, carried out forced labour without any rules or regulations. Accordingly, the number of workers drastically decreased and only 50% remained in this new Group in 1983. Although the organisation survived until 1995, the total number of workers remaining was only 543.[49] In the Voluntary Regeneration Workers Group, as its name implies, institutional control had weakened in the late 1970s. Moreover, the nonchalance in creating a garbage collector's association based on the workers' voluntary participation actually led to the government's abandonment of the Group. As the government stepped back from direct control, the new Group's operations reverted to former patterns, but more replete with internal exploitation.

In the early 1980s, as it had become more difficult to control the workers in one place, the City dispersed them to ten different patches of land, such as behind apartment complexes or in-between high-rise buildings in the formerly undeveloped southern Seoul.[50] In these places, they performed garbage collecting individually or as small groups hidden from view; considering the group-based work patterns of the garbage collectors, the diffusion of the community members into different regions must have dismantled the power of their working society (Figure 2.7). As extensive urban development

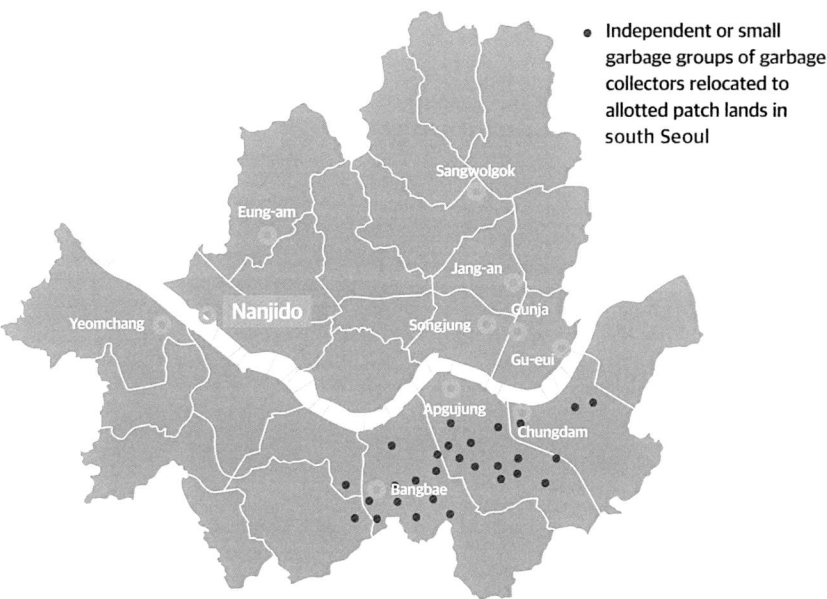

Figure 2.7 Control through non-control: the urban development market system in the early 1980s, the City scattered the garbage collectors to ten different patches of land: for example, in the then underdeveloped southern Seoul. © Jeong Hye Kim

proceeded in southern Seoul in the 1980s, however, the garbage collectors were again forced to leave these allocated lands and move into low-priced leased apartments located far north of or outside Seoul. In these new locations, they could not carry out their jobs, since garbage collecting rested on the consumer market of the urban space[51] (Figure 2.8). By applying legal relocation regulations, corporate developers drove the garbage collectors to farther margins of the city, which disintegrated their living and working community. Accordingly, the national/municipal authority no longer needed to control the garbage workers' presence in the urban space directly, particularly in downtown public spaces. In other words, unlike the previous era when the city government had maintained the cleanliness of the city space through the physical control of ragpicking groups, the City, working in cooperation with private real-estate developers, accomplished the sanitary ideal through urban planning based on the rule of market in this new era.

These changes indicate the transition in the control system of garbage collectors during South Korea's transition from an industrial to a post-industrial era: from the direct physical control of ragpickers (Reconstruction Workers Group) to the tacit control of garbage collectors by way of abandonment (Voluntary Regeneration Workers Group) to the indirect spatial management

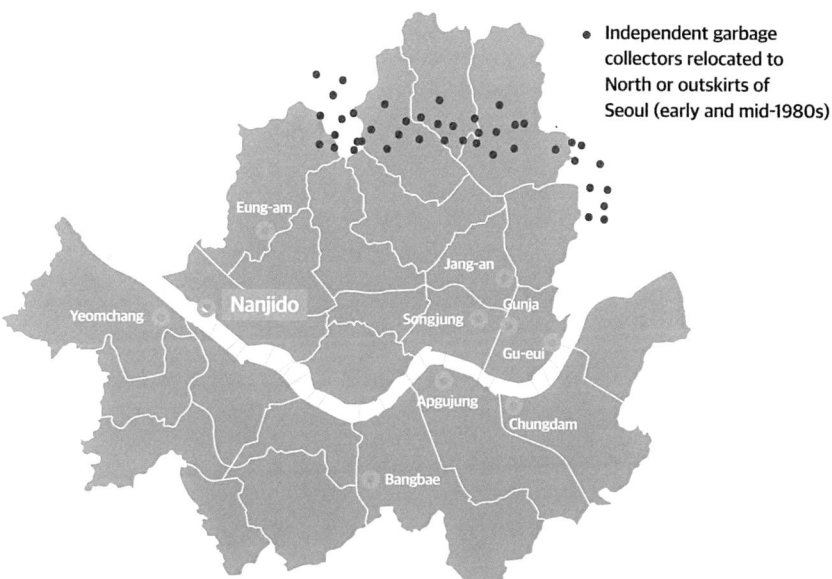

Figure 2.8 Control through non-control: throughout the development of southern Seoul, garbage collectors were forced to move from southern to northern Seoul. Early and mid-1980s. © Jeong Hye Kim

based on real-estate development. The last case ultimately deprived the garbage collectors of their worksite, incapacitating their economic activities.

Meanwhile, the opening of Nanjido Landfill coincided with the transitions in the government's management of garbage collectors in the late 1970s. This gave independent garbage collectors an alternative to their reliance on the government's regulative policies toward waste workers. On the one hand, the real estate market system designated the garbage collectors' residency and worksite, relegating them to the outskirts of the city. On the other hand, the garbage collectors also had a choice to settle within the unprecedentedly massive municipal landfill of Nanjido (Figure 2.9).

As the government began to manage the city's waste in the large-scale municipal landfill, the garbage collectors, who had worked in the city streets, settled in the new landfill area. In short, for the Seoul City government, Nanjido Landfill functioned at once as a topographical waste management system that received Seoul's material waste and as a geo-cultural system that absorbed the socio-economically wasted populations. Essentially, Seoul City did not actively prohibit the garbage collectors' illegal residence in the city-owned site, in part, because Nanjido Landfill rendered the garbage collectors invisible. Their absence in other spaces of Seoul removed their threat to the sanitary state of the urban space.

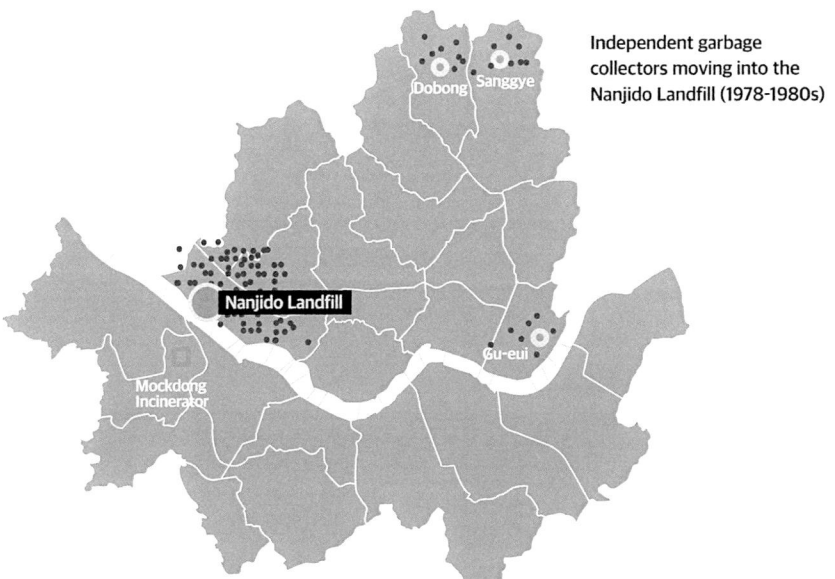

Figure 2.9 Control through non-control (opening of the Municipal Solid Waste Landfill): a large number of independent garbage collectors moved to Nanjido Landfill. Late 1970s–early 1980s. © Jeong Hye Kim

DDT: symbolic sanitary management

In the post-war and postcolonial space of South Korea, the prevention of infectious diseases was one of the government's most crucial jobs as it protected the human resources essential for industrial and military purposes. The active use of DDT (dichlorodiphenyltrichloroethane), from the late 1940s until its ban in 1972, demonstrates the government's vigorous investment in the nationwide preservation of a clean environment and healthy population through epidemic prevention. It was immediately after the nation's liberation from Japan in 1945 and establishment of the US Army Military Government in Korea that Koreans began to use DDT as their main disinfectant. Above all, it is notable that the entire nation regularly practised fumigation with insecticide and pesticide in public spaces for almost 40 years: DDT for almost 30 years and other chemicals for another 10-odd years.

DDT was invented in Switzerland and made available to farmers as an agricultural insecticide. It was also used to eliminate malaria in Europe, North America and Southeast Asia. Malaria is rarely found in Korea since it is geographically located in Northeast Asia; instead, Korea used DDT mostly to kill flies and mosquitoes, the main carriers of infectious diseases like cholera, typhoid, encephalitis and smallpox. Between 1946 and 1972,

DDT had become the most widely used insecticide in this region. It was believed to be so effective that it could eliminate almost all sources of disease in public spaces as well as in private realms.[52] In short, DDT was perceived as the panacea protecting the human body and living environment from any potential pathological threat.

Meanwhile, the chronology of the active use and ban of DDT in South Korea paralleled the prevalent use of and controversies on DDT in the United States. DDT was officially banned in the United States after a long and fierce debate surrounding Rachel Carson's public warning against the potential environmental hazards of DDT (*Silent Spring*, 1962)—more specifically, the debate between chemical manufacturers and their supportive scientists versus unsupportive scientists and the public who demanded DDT be banned for its toxicity and environmental detriment. The history of DDT use in South Korea will elucidate one facet of the West's environmental impact on the post-war and postcolonial space in Northeast Asia, since operations in the West reflected concurrent political situations.

The Korean War and DDT

Korea started to use DDT from 1946 under the US Army Military Government leadership throughout the Korean War and the early Republic of Korea until the chemical's prohibition on 27 December 1972, following the US government's decision to ban its use in June 1972. The US armed forces based around the world heavily influenced the use of DDT, not only in South Korea but also internationally. The US armed forces dispatched throughout the globe during and after WWII decisively contributed to the worldwide distribution and use of DDT. After the US FDA (US Food and Drug Administration) had certified that DDT powder is safe for human beings, the BEPQ (Bureau of Entomology and Plant Quarantine) recommended the armed forces use it as a louse powder and insecticide.

As for the procurement of DDT in South Korea, the US armed forces and international organisations like the UN and UNICEF played a major role in its importation and distribution as an indispensable item for hygienic purposes during the pre-war (between the nation's liberation and the outbreak of the Korean War), wartime and post-war eras. During the late 1940s, the US Army Military Government in Korea imported 150,000 drums of DDT at one time[53] and either sold them to the public at a low price or distributed them to each household for free. People generally considered insecticides and antibiotics crucial in the war and post-war environment, in which people were especially vulnerable to contagious diseases. During the Korean War, the UN provided DDT as one of the relief supplies along with food, clothing and other daily requirements.[54] After the War, apart from the South Korean government's import of DDT, UNICEF (1956) also provided the chemical along with other necessary medicines, such as penicillin.[55] Bags of DDT came from various sources, including large stocks from American manufacturers, which had been in storage for 5 or 6 years.[56] Likewise, during wartime and the early era of national reconstruction, there was an extensive use of DDT in both public and private spaces in South Korea.

Meanwhile, there had been a decades-long controversy over the toxicity and ineffectiveness of DDT. During WWII, particularly from 1942 to 1944, the US Army and American scientists engaged in experiments with DDT to prove its effectiveness as an insecticide and its safety for the human body so that the armed forces could continue to use it to protect soldiers from lice.[57] During the Korean War, armed forces in Korea also conducted research on 'DDT Resistance in Korean Body Lice'.[58] In this research, the authors claimed that while tests on mosquito larvae demonstrated DDT's full insecticidal potency, the routine application of 10% DDT powder on a large group of Korean military personnel during the winter and spring of 1951 resulted only in an increase in the infestation of body lice. The authors pointed out that since 1947, many localities reported DDT resistance in houseflies, and laboratory experiments had demonstrated that selective breeding could develop resistant strains. They also asserted that the effect of DDT on typhus control might be uncertain in some regions.[59] In 'DDT: Its Effect on Fish and Wildlife' (1946), Clarence Cottam and Elmer Higgins demonstrated similar conclusions, noting that 'many insects died within a few days of application of the spray, particularly adult mosquitoes but the effects were only temporary; most species returned to normal numbers in 2 to 3 weeks'.[60] Nevertheless, as with other early research on the potential detriment of DDT in the 1940s, there was no serious reception or public address of the researchers' findings against the effectiveness of DDT.[61]

In South Korea, there had been many reports about the increased number of DDT-resistant insects early in the 1950s. Besides the wartime research, the World Health Organization (WHO), in 1957, also expressed concern that many insects were developing resistance to DDT.[62] From the mid-1960s, announcements and reports on the various opinions about the potential hazards of DDT surfaced in South Korea; moreover, the national decisions of the United States and Japan regarding its use went public.[63] South Korean experts on the natural environment also claimed that DDT might destroy the natural equilibrium by killing some insects that are natural enemies of certain birds.[64] Despite the increasing anxiety over the chemical, without proven toxicity on humans and wildlife within the mainstream field of science in the United States, South Korea continued to use DDT for another decade.[65] However, as more concern for the hazardous potential of DDT arose throughout the mid- to late 1960s in the United States, the national government of South Korea stepped back from its unmitigated reliance on DDT. For instance, it provided the public with an indirect guide to the chemical's use instead of directly fumigating public areas with DDT, and slowly replaced DDT with Dibrom and other insecticides from the late 1960s.[66] Finally in 1972, 6 months after the US government completely banned DDT, the Ministry of Health and Social Affairs of Korea announced its ban on the chemical.[67]

Based on the worldwide distribution and use of DDT and decades of reliance on the chemical during and after the Korean War, this study affirms that the international war situation contributed to the increased concern for a sanitary environment and healthy bodies, especially in post-war, post-colonial countries. Considering that the US armed forces and international organisations mainly distributed DDT to the democratic bloc, we can view the issues of sanitation and health as inseparable from the political situations in the post-war, postcolonial spaces during the latter half of the twentieth century.

A belief system of fumigation

South Korea implemented cleansing, or the creation of a sanitary environment and hygienic conditions for the human body, through DDT in two ways: fumigating public spaces and educating the public on its domestic use. Of the two practices, fumigating public spaces was more significant than the decades of educational campaigns. It is notable that, in some cases, the people demanded fumigation despite the debate on its ineffectiveness and/or harmfulness to human beings and the equilibrium of the environmental ecology.

In the postcolonial space before the Korean War, the US Army Military Government first practised regular fumigation in Korean cities, targeting orphanages and impoverished regions.[68] In the summer and fall of 1949, under the First Republic of Korea, the nation was confronted with an emergency. Fearing an encephalitis epidemic, the government implemented concentrated DDT fumigation all over the cities, while closing elementary schools and cancelling all athletic games in large stadiums.[69] In the following year, on 21 April 1950, 2 months before the Korean War broke out on 25 June 1950, the government designated 24–26 April as a week of cleaning, and continued seasonal fumigations.[70]

During the Korean War, the US Army Air Force major general planned infectious disease control and nationwide DDT fumigation.[71] Based on this plan and the associated manual on fumigation, the government officially implemented seasonal DDT air fumigation across the country.[72] Although the government provided general guidance on domestic cleanliness while practising public fumigation, individual instruction was neither detailed enough nor effective in killing insects immediately.[73] Instead, the government focused on DDT fumigations and citizens continued to demand regular fumigations, too. Likewise, the wartime cleaning activities were based on the immediate effectiveness of removing potential pollutants, particularly through public fumigation, rather than on long-term hygiene and healthcare policies.[74]

After the War, from the 1960s until 1972, the military regime of the Third Republic endeavoured to reconstruct the war-torn country and bring about industrial development by rehabilitating the environment for the healthy productive bodies of the citizens, which justified and further encouraged government-led fumigations and collective campaigns for public sanitation. Above all, the national/municipal government practised DDT air fumigation more often and extensively than before with the cooperation of the army, air force and navy. It especially increased aircraft fumigation in flooded areas and swampy lands after the summer monsoon season. In the summer of 1959, for example, the government implemented air fumigation twice a month in addition to the seasonal fumigation. Instances of public fumigation increased further in the summer of 1964: regularly twice a week or intensively 3 days in a row.[75]

As reliance on DDT escalated during the post-war years, DDT had become a symbolic word in 1960s South Korea; it represented sanitation through the complete human control over the physical environment and embodied the creation of sanitary conditions through the incapacitation of dirt's potency. DDT and its immediate effect on the killing of disease-causing insects, in this context, played the role of an instigator that helped the public to perceive that they could achieve cleanliness, as a condition and ideal of modernisation, by asserting full control over physical dirt (Figure 2.10).

After DDT was officially banned in South Korea in 1972, vaccinations gradually prevented summer epidemics and in the 1980s, the increased availability of vaccines amongst general citizens spurred the transfer of responsibility for sanitation and health from the public sector to the private sector. Nevertheless, the Task Force for Epidemic Prevention remained an honorary position.[76] Seasonal fumigation continued for more than a decade with chemicals other than DDT, selectively in impoverished areas, including Nanjido Landfill, which was not properly equipped with infrastructure like a sewage system.[77]

Essentially, the magnitude of fumigation with chemicals was a reflection of the public's belief in the practice. For the post-war generation of South Korea, who had depended on DDT fumigations for a sanitary environment and healthy bodies for decades, the cloud of powder, released by means of manual devices, vehicles or aircrafts, was the symbol of cleanliness and safety from disease. The belief in the invisible and/or intangible characteristics of most of the fumes was fundamentally rooted in the belief system of modern Western science; the advanced science, represented by fumes, would purportedly protect them from germs, symbolised by dirt, which they believed would bring crisis to their lives. Populations in war-torn environments also perceived DDT as a physical-mental protection system associated with the ideological trust in or dependency on anti-communist democracy. The massive fumigation in public spaces especially provided the people with psychological relief, giving them the reassurance that both the

Figure 2.10 DDT fumigation by military vehicles in downtown Seoul. 1958. ©
National Archives of Korea

chemical (modern science) and the city or nation (modern social system)
protected them.

In summary, from the late 1950s until the end of 1972, the period when
DDT was legally the main insecticide, cleaning was executed in a strikingly
collective way. Throughout the history of post-war Seoul, the government
has not managed any other chemical with as much intensity or totalitarian
direction as it had DDT. This is in part because the worldwide distribution
of DDT coincided with the political circumstances and governing systems
of the postcolonial and post-war space of South Korea. The nation strived
to control the chaotic socio-political situation and construct an orderly
social structure based on the notion of sanitation. All the same, the public's
traumatic psychological insecurity and fear of a potential physical threat
actually operated the governments' sanitary policies and consolidated the
system of belief based on modern science and the modern social system. In
this sense, the nationwide and city-wide fumigation of public spaces had as
much symbolic meaning as physical impact.

Meanwhile, as government control over the garbage collectors weakened and was handed over to the urban development market system from the mid- to late 1970s, the management of epidemics and general hygiene, which the government had primarily overseen, was gradually transferred from public control to private responsibility during this period. With the shift from an industrial to a post-industrial era, the prevailing economic logic started to overshadow the link between sanitation and morality and/or ideology, which began to lose potency. However, we must distinguish the predominance of the logic of the free market economy over institutional control from political democratisation grounded on social equality. The developed capitalist market system legitimately removed the social safety net for people who could not become proper producers and consumers, and made them invisible in the urban space; this was the market economy's sanitisation of urban space.

Nanjido Landfill: spatial sanitary management

The waste management sites and waste itself

There were dozens of unofficial garbage dumping sites in various areas of Seoul in the 1960s and 1970s, but they mostly existed within residential and commercial sites. It was not until the official opening of the Municipal Solid Waste Landfill of Nanjido in 1978 that non-waste and waste—appropriate and inappropriate objects categorised by the norms of a society—were explicitly divided within the urban space. Although Seoul City had formerly disposed of material waste in the waste dumping sites around the city, it next attempted to separate the useful and the useless completely, especially during its transition into a consumerist society, by prohibiting the mixture of valueless objects with those of value. That is, as the notion of sanitation developed a closer connection to issues of income class in the post-industrial economic situation, zoning more significantly reflected the city's tendency to maintain sanitation through the dichotomous division of urban space into clean/appropriate and dirty/inappropriate realms. Likewise, at the onset of the 1980s, relegating various substances to the right realm based on their in/appropriateness as measured by the socio-economic criteria of the time was essential for the sake of the cleanliness, orderliness and wellbeing of Seoul's urban space. On the one hand, from this point of view, Nanjido Landfill was a waste management site, or a spatially rendered sanitary zone that facilitates the sanitation of the urban space of Seoul. On the other hand, it was a wasted entity, or dirt itself, which constantly polluted and threatened the cleanliness, orderliness and wellbeing of the city.

As a waste management site, Nanjido Landfill had received most of the municipal solid waste produced in Seoul for 15 years. From a quantitative perspective, Nanjido Landfill opened as an unprecedentedly large-scale municipal landfill to receive the exponential amounts of waste produced from the increases in consumption. Since its opening, Nanjido Landfill had mainly operated in two sites (Landfills 1 and 2). Operations also took place in the cargo and sludge-dumping site in the east (Landfill 3) and in some areas north of Landfills 1 and 2. During the early period, approximately 1,000 eight-tonne trucks dumped waste daily. This extensive dumping filled about 16.5 m² of the wetland and eventually levelled the ground by filling Landfill 1 in 1980 and Landfill 2 in 1985. Since overground dumping had begun in 1986, an average of 3,000 eight-tonne trucks of waste came into Nanjido every day, raising the height of the landfill by 4 m per year. The average daily amount of waste significantly increased from 27,116 tonnes in 1987 to 28,877 tonnes in 1988.[78] By the time of the landfill's closure in 1992, the total scale of the landfill reached 92,000,000 m²: 56,400,000 m² in Landfill 1, 34,800,000 m² in Landfill 2 and 800,000 m² in Landfill 3 (the sludge-dumping site). The total amount of waste dumped in Nanjido over 15 years of operation was an estimated 101,120,000 tonnes.

From a qualitative perspective, the type of waste in Nanjido Landfill also changed as the industrial patterns transformed into more consumerist post-industrial ones. In the late 1970s and early 1980s, used coal briquettes had amounted to about 80% of the total waste (as of 1978). However, the proportion dramatically decreased by 51.70% in the mid-1980s (as of 1986) and vinyl, papers and fabrics replaced coal as the major type of waste. Correspondingly, garbage collectors largely recycled papers and fabrics along with plastic and metal. According to a study conducted on the landfill's pollution upon its closure, combustible waste amounted to 41.4% and 18.8% (non-combustible 58.6% and 81.2%) in Landfills 1 and 2 respectively; the high portion of combustible waste in Landfill 1 corresponded to the illegal dumping of industrial waste in this site. The average carbon content reached 14.15%, which could mainly be attributed to the increased amount of vinyl and plastic since the mid- to late 1980s. It led to anaerobic decomposition, slowing down the speed of the earth's natural regeneration. What's more, Nanjido Landfill functioned as a spatially arranged sanitary management facility, which received and treated waste through diverse methods of land-fill creation, e.g. top-soiling.

On the other hand, as the site received a massive amount of waste, the landfill itself had become a structure of waste. Nanjido Landfill was rife with pollutants because it was an unsanitary landfill, where all types of waste were dumped without pollutant treatment systems, and the waste of various non-decomposable materials rapidly increased from the mid- to late 1980s. This issue calls the architectural and landscape identity of the landfill into question. An unsanitary landfill is at once a human-made structure and a mass of polluted materials that gradually merges with the natural environment. In a discussion of sub-natural elements, David Gissen illustrates the miasmatic gas in a similar sense: 'It is a form of human-produced nature that is imagined to impart both new forms of environmental and social degradation'.[79] Both the natural environment and society create the landfill as human-produced nature, and that landscape in turn influences both the natural environment and society. In this sense, the landfill is a metaphoric structure that represents the relational aspect of the natural environment and the social dynamics of the time.

Fumes and borders

•   Battle of fumes

Throughout Nanjido's landfill period (1978–1992) and until the completion of the landfill's regeneration in 2002, Seoul City controlled the harmful insects and miasmatic odours coming out of the landfill site with fumigation. Unlike in the previous era, the City used fumigation in two ways: one for insecticide, and the other for deodorisation, or the removal of gaseous odours. To that end, the City practised regular fumigations in the landfill since Nanjido Landfill's opening until its closure; during the mid-1980s, the government sprayed agricultural pesticides and/or deodorisers from vehicles or helicopters two to three times weekly.[80]

The fumigation practice during the landfill period had several notable characteristics; first, its fumigation was highly concerned with deodorisation; second, the city-wide spraying of the past had narrowed down to the concentrated zone of the landfill; third, the landfill's fumigation was essentially for the neighbouring regions' cleanliness and wellbeing. Ultimately, the City used fumigation to hide the waste and Nanjido Landfill itself, making it separate and unrecognisable from the urban space.

In fact, flies and other insects were the immediate causes of inconvenience for the residents of Nanjido Landfill and North Sangam-dong town. As mentioned previously, however, the City focused more on deodorisation because the stench negatively affected the neighbouring regions and the general image of the region and the city.[81] During international events held in Seoul, the municipal government was especially concerned with the miasma around the landfill site. In 1988, as the Seoul Olympic Games grew near, the City increased the frequency of spraying deodoriser into the air to three times a week.[82] In contrast, complaints about unpleasant odours were not as serious 14 years later, around the time of the FIFA 2002 World Cup Korea/Japan and the landfill's regeneration into a park; yet, the city government was still very sensitive about the unpleasant smell from the former landfill site, and promptly responded to any complaints about it.[83]

The gaseous odours were unlike other pollutants produced from the waste dump. For example, the leachate that polluted the stream between the landfill area and North Sangam-dong town created a clear borderline between the two realms. Whereas, the gaseous odours not only functioned as the mark of a site of dirt, but also as a threat to the sanitation of other regions since they diffuse unpredictably. The landfill odours were so potent and uncontrollable that the site's managers could not completely prevent them from spreading beyond the landfill's boundaries. As a result, the government consistently practised regular fumigations to inhibit the odours from crossing the border between the two separate zones of the city—the zone of the valuable and that of the valueless. Due to the constant production and control of odours, Nanjido Landfill was a battleground of fumes: the miasma produced by the landfill and the chemical fumes sprayed to kill the landfill's putrid odours.[84]

- Border and odour

To treat the city-produced waste efficiently, most landfills occupy a certain portion of the city, addressing the cleanliness, orderliness and wellbeing of the other parts of the urban space. In this respect, the landfill, located within its own particular urban space, is essentially an internal colony seized by its own governing body, and local sanitation practices are acts that relocate supposedly inappropriate objects to that colonised territory. During this process, bordering devices emerge and the politics of exclusion prevail. In this context, we can recognise Mary Douglas's account of 'dirt'[85] as equivalent to the idea of waste, and 'purity' as commensurate with non-waste—the state that fits the sanitary criteria of a given society. Douglas's discourse on 'dirt as matter out of place' is particularly notable here as she addresses the ideas of dirt, purity and orderliness in spatial terms. This provides the foundation for a discussion on the meaning of Nanjido Landfill as a spatially constructed sanitary management device within the urban space of Seoul.[86]

Drawing on Douglas's argument on purity and danger, Bauman defines design as the constant process of dividing the material outcomes of an action into 'what counts' and 'what does not count', or into 'useful objects' and 'waste'.[87] With explicit propositions, such as 'modernity is a condition of compulsive and addictive designing' and 'here is design, there is waste', he affirms the relationship between design and the production of waste. Taking a step further, Bauman applies this idea to the social design that categorises human beings into appropriate and inappropriate members of the social space:

> When it comes to designing the forms of human togetherness, the waste is human beings. Some human beings who do not fit into the designed form nor can be fitted into it. Or such as adulterate its purity and so becloud its transparency [...] oddities, miscreants, hybrids who call the bluff of ostensibly inclusive/exclusive categories.[88]

Seen from the dichotomous conception of modern design, the landfill, as a structure, receives all objects categorised as inappropriate by the norms of the urban space. Meanwhile, the landfill, as waste itself, must also be expelled to the boundaries of the urban space: both the physical boundary and the social one that reflected the urban imaginaries. Accordingly, the

City must control the landfill's pollutants, or dirt, that damages the earth, water and air conditions and prohibit them from permeating into and contaminating the clean and orderly urban space.

In Nanjido Landfill's case, one of the major elements that distinguished the landfill site from other regions was the miasma emitted from the rotting mixture of the unsanitary landfill. The malodorous smell from Nanjido Landfill created the distinct border that separated the landfill from non-landfill regions. It even superimposed an alienated image on the landfill area. An anonymous interviewee of this research claimed that, when s/he was riding a bus that terminates within the landfill region, s/he immediately recognised that s/he had crossed the border and entered the Nanjido Landfill area by the miasma. Several former Nanjido Landfill residents who partook in the interview also remembered the foul odour as one of the most significant factors stigmatising the landfill region:

> You may not be able to imagine how obnoxious the odours of Nanjido Landfill were. As time went by, however, we became anosmic and didn't even sense the foul smell. The problem is that the smell seemed to soak into our bodies, our body cells. When sweating in the summer, the same putrid smell emanated from our bodies.[89]

> Compared to other people from Nanjido Landfill, I more often went outside the landfill to meet friends. Whenever I visited my friend, her mother used to say to me, 'Go! Go! You stink!'[90]

In a study on the stench of the poor in nineteenth-century European society, Alain Corbin explains the history of the shift in factors determining stench from the topography and natural phenomena to the social circumstances and class. Amongst the various groups of poverty—mainly ragpickers, sailors and factory workers—he specifically mentions ragpickers as the archetype of stench, calling their smell the fetidity of the labouring classes, or the 'secretion of poverty'.[91] Corbin accounts for the personal odours of the working class and points out that perspiration, when all other smells of excreta have been eliminated, reveals the inner identity of 'I', the individual. At that time, the bourgeois had become increasingly sensitive to olfactory encounters that conveyed the disturbing aspects of intimate life. Corbin asserts that, in part, the tendency of the bourgeois class to project onto the poor what they were trying to repress in themselves can explain the emergence of individual smell.[92] In addition, sanitary reformers established a link between stench and the anosmia of the masses (e.g. sailors are keen-eyed but do not have a good sense of smell nor hearing), and that link supported the bourgeoisie in their belief in the odour of the poor and the need for deodorisation.[93]

Contrarily, however, Corbin places more emphasis on the eighteenth-century anthropologists' discovery that stench is related to climate, diet, profession and temperament rather than to wealth or poverty.[94] Corbin's argument implies that it is more reasonable to link stench to the character-istics of an individual's profession and habitational environment than to view bodily odour simply as an issue of class. His studies demonstrate that there is no concrete justification for the relationship between fetid stench and poverty as it was an invention of the bourgeoisie. The same was true of the Nanjido Landfill community; the division between the landfill area and other regions did not entirely rely on sensory (olfactory) elements, but also depended on socio-psychological factors, or the preconception that identi-fied the garbage collectors in the landfill with the waste, or dirt itself.

Although odours had created a border, isolating Nanjido Landfill from the other regions of Seoul, such a border did not have a clear demarcation because odours do not stay within a clear borderline; they unpredictably cross any visible and/or invisible boundaries. The threat of gaseous odours stems not only from the idea that they possibly retain the causes of disease or pestilence, but also from their very essence of ambiguity, uncertainty and unpredictability, which modern society has incessantly resisted. While building a border by separating the realm of waste from that of non-waste but freely crossing the borderline of its own creation, the odours of the landfill allude to the subversive potential of re-mixing pre-categorised waste and non-waste, which would ultimately disrupt the current dichotomous standards of order.

Notes

1 To distinguish individuals who had been subordinated to private ragpicking groups or independent ragpicking before 1961 from individuals who had been registered on the government list (after 1 July 1961), I call the former 'rag-pickers' (*nung-ma-ju-i*) and the latter 'garbage collectors' (*pye-poom-su-jip-in*) because 'garbage collector' was the official title that Seoul City government had attributed to registered ragpickers. Since the Korean term *pye-poom-su-jip-in* literally means the collector of wasted yet recyclable items, I use the same title of 'garbage collector' for the Nanjido Landfill waste workers who sorted the recyclables from the mixed garbage dump.

2 Lorraine Code claims that Rachel Carson's work prefigured, if sometimes tac-itly, a subversiveness challenging entrenched power structures and a tenacious belief in the 'natural' (pre-1960s) order of things, both human and nonhuman. Code also contends that Carson is one of the most eminent precursors of the new social movements fuelled by the 1960s civil rights and women's movements in North America; environmental and antimilitarist movements; anti-imperialism, postcolonialism and the hegemony of Enlightenment rationality in post-1968 Europe (Code, 2006: 14–20, 47).

3 Mary Douglas's concept of 'dirt' has been used by diverse scholarly approaches, such as structuralism and phenomenology, which created tensions between dif-ferent interpretations of 'dirt' as symbolic or material. In the examination of the landfill, viewing 'dirt' from both material and symbolic perspectives is inevitable because the landfill is a physical mass of dirt that implicates how people render and operate apparatuses of modernity and [post-] industrial society.

 4 Zygmunt Bauman, 'Dream of Purity', *Theoria: A Journal of Social and Political Theory*, No. 86. Dimensions of Democracy, October 1995: 51; Douglas, 2002 [1966].
 5 I use the term 'modernisation' to indicate the state of a society that is supported by its own industrialisation and accompanying urbanisation. I view the post-war era of Seoul, South Korea in the 1960s and 1970s as a space in the process of industrialisation and modernisation, with ideal imaginaries of the industrialised and modernised urban space. The use of the term 'modern' is limited in the discussion of the pre-war era, including the colonial period, as the state of a society that operates by the logic of westernised science and reason.
 6 The US Army Military Government in Korea (1945–1948), the Korean War (1950–1953), the First Republic of Korea (1948–1952, 1952–1960), the Second Republic of Korea (1960–1961), Military Coup d'état (1961), military dictatorial regimes (1963–1987) and transitional administration from military dictatorship to democratic government (1988–1992).
 7 *Dong-A Ilbo* (9 July 1959) and *Kyunghyang Shinmun* (12 June 1961). The national and municipal hygienic guidelines, provided since the late 1950s, were effective until the 1970s.
 8 Newspapers reported seasonally on general safety guidelines and the use of DDT after the Korean War until 1972. See *Kyunghyang Shinmun* (19 Jan 1956–24 Aug 1958; 1 May 1961–25 Jan 1964; 27 Jul 1966–15 Jul 1968), *Dong-A Ilbo* (27 Jun–19 Sep 1954; 12 Jun 1956–11 Jul 1959; 23 Jun 1961–16 Jul 1962; 8 Mar 1965; 26 Jun 1969) and *MK Business News* (18 Jun–16 Jul 1968).
 9 *Kyunghyang Shinmun* (21 August 1970).
10 *Dong-A Ilbo* (15 July 1974).
11 It is notable that the mayor of Seoul, Taek-shik Yang (1970–1974), emphasised sufficient water supply and the sanitation of crowded public spaces as primary conditions for clean and safe holidays in his speech delivered on the Chuseok holiday (*Dong-A Ilbo* [20 August, 14 September 1970]).
12 *Kyunghyang* Shinmun (11 August 1975).
13 The textbooks include *The Textbook on Moral Training for Junior High School* (Seoul: Hwimungwan, September 1906), *The Textbook on Moral Training for High School* (Seoul: Hwimungwan, August 1907) and *The Textbook on Moral Training for Elementary School* (Seoul: Dongmunsa, July 1909). The Japanese administration prohibited the publication of *The Textbook for Junior High School* in 1909. In 1977, the Korea Research Centre for Bibliography republished a photoprint copy of the book.
14 'Su-shin' means 'training', 'discipline' and 'cleaning of the body'. In the sense of tradition (Confucianism of Joseon Dynasty), 'su-shin' emphasises the awakening of oneself and self-reflection, whereas 'su-shin' in *The Textbook* is concerned with the 'morality' required for modern citizens. The transfiguration of the body and nurturing of a healthy body were indispensable to *The Textbook* (Cheol-woon Kim, 'Modern Transfiguration of Moral Training: The Individual Confined to the State', *A Collection of Philosophy Treatises*, Vol. 2, No. 48, 2007: 137). Herbert Spencer's Social Darwinism and J. K. Bluntschli's idea of the nation-state as an organic system form the foundation for this idea. Whereas Japan deployed Social Darwinism as the foundation for their imperialistic nation-state, Korea used these notions in an effort for the nation to overcome imperialism (Cheol-woon Kim, 2007: 140).
15 Ibid.; see *The Textbook on Moral Training for Elementary School*: 1–19.
16 Ibid.: 148–149; see *The Textbook on Moral Training for Junior High School*, chapter 1 titled 'Taking care of life', section 8 on 'health': 8.

17 The Japanese colonial administration carried out hygienic policies that the West had previously undertaken. Regarding the hygiene police of the West, see Alain Corbin, *The Foul and the Fragrant: Odour and the French Social Imagination*, London: Picador, 1994: 160; A. A. Mille, 'Rapport sur la mode d'assainissement', 1854: 223.

18 Corbin points out that Durkheim insisted on distinguishing the moral element from society's preoccupation with hygiene, but before him the moral implications of public health ventures were emphasised on many occasions in the seventeenth and eighteenth centuries. See Corbin, 1994: 157; Luc Boltanski, *Prime education et morale de classe*, 1969: 110.

19 'Welfare nation' at the time meant an industrially developed and economically wealthy nation, which is different from the contemporary meaning of a national system that promotes the socio-economic wellbeing of its citizens based on equality of opportunity.

20 Another perspective contends that this movement inherited the Joseon Farming Village Development Movement (1931–1940), which the Japanese colonial government had initiated. The Reconstruction Citizens Movement is regarded as the precedent for the New Town Movement (1970–) of subsequent decades. See Soo-gul Ji, 'Japanese Militarism, Fascism and Joseon Farming Village Development Movement', *History Criticism*, Summer 1999 and Jin-do Park, Do-hyeon Han, 'New Town Movement and Yushin Regime (the Fourth Republic): On Park Chung-hee Regime's Farming Area New Town Movement', *History Criticism*, Summer 1999.

21 The first head of the Office, Jin-oh Yoo, regarded freedom and social control as complementary, and defined freedom in the modern nation as '[freedom] granted by national authority'. For him, freedom can be achieved within the frame of anti-communism and disciplinary governance (Eun Huh, 'Reconstruction Citizens Movement during the May 16 Military Regime', *Historical Research*, No. 11 (December 2003): 16; Jin-oh Yoo, 'Freedom and Power: One Logic', *Chosun Ilbo* (August 1955). The second head of the Office, Dal-young Yoo, claimed three doctrines: 1) autonomy through the individuals' internal discipline, 2) modernisation based on tradition and 3) enlightenment of citizens based on anti-communist ideology. See Eun Huh, 2003: 17–18; Dal-young Yoo, 'Revolutionising People', *Discovery of Human*, Seoul: Eomungak, 1963.

22 The third head of the Office, Gwan-goo Lee (May 1963–February 1964), affirmed independent unification based on anti-communism in his principles of the Movement. See Eun Huh, 2003: 22; Gwan-goo Lee, 'May 16 Coup and Unification', *Supreme Council*, No. 1, August 1961.

23 The government collaborated with Christian organisations to find an alternative ideology that could stand against both communism and extremist liberal democracy (Eun Huh, 2003: 22–23, 30).

24 In October 1972, President Park Chung-hee declared an emergency state of siege; by dispersing the National Assembly and prohibiting political party activities, he took complete authority over legislation, jurisdiction and administration. Regarding his initial intervention in the epidemic prevention policies, see *MK Business News* (3 April 1970).

25 *Kyunghyang Shinmun* (30 July 1971).

26 *Kyunghyang Shinmun* (2 September 1970).

27 *MK Business News* (15 August 1970).

28 *MK Business News* (10 March 1972).

29 *MK Business News* (6 February 1975).

30 *Kyunghyang Shinmun* (28 March 1970).

31 Seung-Hee Kim, 'The Bio-political Power and Its Limitations Seen through the Cholera Epidemic in 1969 in South Korea', *Social Thoughts and Culture*, Vol. 18, No. 1 (March 2015): 225–226.

32 Regarding the details of this incident, see Seung-Hee Kim, 2015; *Kyunghyang Shinmun* (3, 4 September 1969).

33 Ibid.: 226; Jong-hui Jeon, 'Chronological Changes of the Epidemics before and after the Korean War', *Infection*, Vol. 2, No. 1, 1970: 17–22.

34 Ibid.: 234–235.

35 Since some ragpickers were homeless for bad behaviours, ragpickers were also called *yang-a-chi*. It caused confusion in distinguishing those who were ragpicking as a profession from delinquent homeless teenagers. See Soo-jong Yoon, 'Ragpickers and Nation', *The Radical Review*, No. 56, Summer 2003b: 265–268; *Dong-A Ilbo* (14 February 1961).

36 Soo-jong Yoon, 2003b: 269; *Dong-A Ilbo* (6 February 1961).

37 *nung-ma-gong-dong-che*, No. 5, 1991: 39; *Dong-A Ilbo* (14 February 1961).

38 After the May 16 coup (1961), the military powers established the Central Information Agency (*jung-ang-jeong-bo-bu*) as the authoritative governing entity, and the National Reconstruction Supreme Council (*guk-ga-jae-geon-choe-go-hoe-ui* established on 20 May 1961) to administer state affairs.

39 Soo-jong Yoon, 2003b: 270; Sung-soo Bang, *The History of Gangsters*, Seoul: Salim, 2003.

40 Ibid.: 274; *Kyunghyang Shinmun, Dong-A Ilbo, Chosun Ilbo* (2 May 1962) and *Chosun* Ilbo (14 May 1962).

41 Ibid.: 274–275; *Chosun Ilbo* (15 May 1962) and *Kyunghyang Shinmun* (14 May 1982).

42 Ibid.: 277–278; Buja Im, 'Special Missionary Work for Working Teenagers: On Reconstruction Workers', MA thesis at Berean Christian School, 1976: 228.

43 Ibid.: 275; *Kyunghyang Shinmun* (30 June 1962).

44 See the chapters on 'Docile bodies' and the 'Means of correct training' in Michel Foucault, *Discipline and Punishment*, trans. by Alan Sheridan, New York: Vintage Books, 1995 [1977].

45 Somewhat similar patterns of garbage collecting and trading processes continued in Nanjido Landfill in the 1980s and 1990s.

46 Soo-jong Yoon, 2003b: 278; *Dong-A Ilbo* (6 March 1974).

47 In 1975, as government control relaxed, the Reconstruction Workers Group (renamed the Korean Workers Reconstruction and Welfare *han-guk-geun-ro-jae-geon-bok-ji-hoe*), was handed over to private institutions such as the Korean Council of Retired Policemen and religious organisations such as the Korean Ant Associates (*gae-mi-hoe*, established in 1965). The private groups reinforced the existing exploitation systems. As for the Korean Council of Retired Policemen (*dae-han-min-guk jae-hyang-geong-woo-hoe*), see *The History of The Korean Council of Retired Policemen* (2008), *Chosun Ilbo* (18 January 1976, 16 October 1977) and *Kyunghyang Shinmun* (8 February 1978).

48 Ibid.: 292.

49 Ibid.: 290–292; National Police Agency, 'Abolition of the Voluntary Workers Group', 1995: 3.

50 Independent garbage collectors who had been scattered throughout southern Seoul in the 1980s created their own *nung-ma* community (*nung-ma-gong-dong-che*), but it could not play the role of an organisation or union because of the workers' disconnected residences. See Soo-jong Yoon, 'Characteristics and Changes of Ragpickers Community', *Democracy and Human Rights*, Vol. 2, No. 1, 2003a: 175–212.

51 Soo-jong Yoon, 2003b: 293.
52 Historian John Perkins noted that Paul Herman Müller's research in 1948 was guided by seven criteria for a 'desirable commercial product', which included 'zero or weak toxicity toward warm-blooded animals' (John Perkins, 'Reshaping Technology in Wartime', *Technology and Culture*, Vol. 19, No. 2, April 1978: 171). See also Davis, 2014: 39–40; Paul Hermann Müller, *Histoire du DDT*, 1948. The wide use of DDT was in part based on the lack of clear evidence that it was harmful to warm-blooded animals, including human beings.
53 *Dong-A Ilbo* (18 May 1948).
54 On the UN relief supplies, including DDT, imported to Korea during the Korean War, see *Dong-A Ilbo* (11 October 1950; 11 January, 25 September 1951).
55 On the import and distribution of DDT in South Korea after the Korean War, see *Kyunghyang Shinmun* (11, 24 June 1954; 7 June 1956–14 Dec 1958; 11 Nov 1960; 18 Nov 1967; 24 June 1968), *Dong-A Ilbo* (29 Jul, 14 Sep 1954; 22 May, 16 Oct 1957; 23 Nov 1966).
56 Herbert Hurlbut, Robert Altman, Carlyle Nible, 'DDT Resistance in Korean Body Lice', *Science (AAAS)*, New Series, Vol. 115, No. 2975 (January 4, 1952): 11.
57 Many experts conducted research and tests to prove the insecticides' safety or potential harmfulness to the human body during wartime. Raymond Bushland and his staff from the Sanitary Corps of the US Army developed methods of testing various insecticides, including DDT, on body lice (Davis, 2014: 41). One British researcher also conducted research on the effects of DDT on human subjects. After experiments with 50 volunteers wearing undergarments saturated with 1% DDT, he found no manifested symptoms suggestive of toxic absorption and concluded that soldiers under battle conditions could safely wear garments saturated with DDT as a deterrent to lice (Davis, 2014: 66–67; G. R. Cameron, 'Risks to Man and Animals from the Use of 2,2- bis [p-Chlorophenyl], 1,1,1,- Trichlorethane [DDT]', *British Medical Bulletin*, No. 3, 1945: 234). Such partial experiments contributed to the decades-long use of DDT worldwide.
58 The research article was published in *Science* (1952). The co-authors Herbert Hurlbut, Robert Altman and Carlyle Nibley served the Fleet Epidemic Disease Control Unit No. 1 and 37th Preventive Medicine Company, Eight US Army, Korea (EUSAK).
59 Herbert Hurlbut et al., 1952: 11–12; W. V. King, J. B. Gahan, *Journal of Economic Entomology*, No. 42, 1949: 405.
60 Davis, 2014: 59; Clarence Cottam and Elmer Higgins, 'DDT: Its Effect on Fish and Wildlife', *U.S. Department of the Interior, Fish and Wildlife Service Circular*, No. 11, 1946: 1.
61 During and immediately after WWII, DDT underwent extensive scrutiny by scientists from public organisations, including the USDA, PHS, FDA and the US Fish and Wildlife Service (FWS), as well as from research universities, agrochemical companies and private institutions. However, researchers then hesitated to make broader conclusions from individual tests: e.g. what specific changes might affect other organisms or the ecosystem as a whole. For early research on DDT in the United States and the UK, see Davis, 2014: chapter 2.
62 *Kyunghyang Shinmun* and *Dong-A Ilbo* (11 August 1957).
63 For example, the WHO warned against the outdoor usage of DDT, fearing the extinction of freshwater fish, and health specialists explored DDT as a carcinogen in an international congress on cancer. See 'Wary of Extinction of Freshwater Fish and Birds', *MK Business News* (10 March 1971) and 'Ways to Conquer the Cancer', *Kyunghyang Shinmun* (2 November 1966).
64 'Birds in Danger of Extinction', *Kyunghyang Shinmun* (21 March 1968).

65 In South Korea, although people were generally aware of insects' resistance to DDT, they continued to believe that it was effective in the control of disease anyway. See *Dong-A Ilbo* (23 June 1961).
66 On the warning against or banning of DDT after the Korean War until 1972, see *Kyunghyang Shinmun* (2 Nov 1966; 21 Mar 1968–13 Dec 1969; 15 June 1972), *Dong-A Ilbo* (21 Nov 1969; 15 June, 27 Dec 1972), *MK Business News* (10, 22 Mar 1971; 27 Dec 1972).
67 *Kyunghyang Shinmun* and *Dong-A Ilbo* (15 June 1972); *Kyunghyang Shinmun* (27 December 1972); US EPA, 'DDT Regulatory History', *EPA report* (July 1975). www.epa.gov/aboutepa/ddt-regulatory-history-brief-survey-1975 (accessed on 1 February 2016).
68 *Dong-A Ilbo* (18 May 1948).
69 *Dong-A Ilbo* (19 August 1949).
70 *Kyunghyang Shinmun* (21 April 1950).
71 The South Korean government awarded the US army major general, known as Hue, the Order of Military Merit, and he received citizenship from the Seoul City government for these services. See *Dong-A Ilbo* (29 September 1951).
72 *Dong-A Ilbo* (17 December 1950–9 September 1952), *Kyunghyang Shinmun* (28 June 1952–4 August 1953).
73 During these times, the US Army CAC (Combined Arms Centre) based in South Korea provided the hygiene department managers of each province with professional education on the maintenance of a wholesome environment, including instruction on DDT use. In the following decades, autonomous city governments, especially the capital city of Seoul, took full charge of their region's hygiene. See *Dong-A Ilbo* (1 October 1951).
74 The new insecticide 'lindane' was also introduced as several times stronger than DDT. See *Dong-A Ilbo* (25 November 1952).
75 On DDT fumigation after the Korean War until 1972, see *Kyunghyang Shinmun* (5 June 1954–3 Sep 1958; 27 June, 29 Aug 1962; 5 June 1964; 22 Aug 1966; 24 Aug 1970), *Dong-A Ilbo* (14 Jul 1955–15 Aug 1957; 15 Feb–26 Aug 1959; 3 Aug 1960; 25 Jun 1961–22 Jun 1964).
76 The government made a new position of the honorary supervisor of epidemic prevention in 1985 (*Kyunghyang Shinmun* [14 August 1985]).
77 The government continued to set epidemic prevention policies for the summer, including authorising emergency ordinances and organising task force teams until the mid-1980s (*Kyunghyang Shinmun* [28 April; 2 June 1986]), *MK Business News* [23 July 1985; 16 June 1986]).
78 The increase in waste galvanised Seoul City to prepare an alternative landfill site outside the city to treat Seoul's waste and that of satellite cities while discussing the closure of Nanjido Landfill. The City also built an incinerator in Mok-dong to treat 150 tonnes of waste per day (Seoul City—Park and Management Office, *Making of the World Cup Park*, August 2003b: 14).
79 David Gissen, *Subnature*, New York: Princeton Architectural Press, 2009: 61.
80 *Dong-A Ilbo* (23 July 1984).
81 The indigenous Sangam-dong residents to the north of the landfill petitioned the City to lessen their suffering from rampant infestations of flies and other contagious insects. For the difficulties in their daily lives, see Seo-young Park, 'Historical Transformations of Sangam-dong based on the Lived Experience of Its Local Residents', Master's thesis, Yeonsei University, Seoul: 2004.
82 *Kyunghyang Shinmun* (2 June 1988), *Dong-A Ilbo* (3 August 1988).
83 In 1999, when a citizen issued a complaint about the odour around the World Cup stadium, Seoul City promptly responded to it (*Kyunghyang Shinmun* [17 September, 1 October 1999]). In a city-commissioned research report

titled *Evaluation of Nanjido Landfill and Environmentally-friendly Restoring Strategies* (Seoul Institute, 2000), the researchers particularly emphasised the need to control the odour before the World Cup games, proposing high-pressure spraying of deodoriser if the stench persisted after building the gas treatment facility (Seoul Institute, 2000: 254–257).

84 Gissen points out that in the case of European cities, a new sensitivity to miasma coincided with transformations in urban spaces, such as the distancing of stables and cesspools away from everyday life. People feared the strong odours emanating from human waste were carriers of disease and pestilence. Meanwhile, Gissen asserts that the sewerage systems, established as a solution to odours, actually produced new forms of gas and odours (Gissen, 2009: 58–59).

85 Although this phrase is often attributed to Douglas and the first edition of her account *Purity and Danger* (1966), in later editions Douglas referenced it to Lord Chesterfield—Philip Dormer Stanhope, 4th Earl of Chesterfield, 1694–1773—without providing an exact source. Based on Douglas' biographer Richard Fardon's evidence, Ben Campkin clarifies that Douglas wrote in her notes from her second field trip to the Belgian Congo in 1953, 'Dirt is any matter displaced' (Ben Campkin, 'Placing "Matter Out of Place"', *Architectural Theory Review*, Vol. 18, No. 1, 2013: 47–48).

86 For the spatial interpretation of Douglas's ideas of 'dirt' and 'purity', see Campkin (2013).

87 Zygmunt Bauman, 2012 [2004]: 25.

88 Ibid.: 30.

89 Interview with Nun Magdalena (26 August 2014).

90 Interview with Jae-Soon Yoo (21 August 2014), novelist and journalist, who lived in Nanjido Landfill for several years in the early 1980s.

91 Corbin, 1994: 142–160.

92 Ibid.: 143.

93 Ibid.: 147.

94 Ibid.: 144.

3 Nanjido Landfill as human habitat

Nanjido Landfill was an inhabited landfill where people gathered and settled down to find economic opportunities, eventually creating the garbage collectors' community of the largest scale in the history of Seoul, South Korea. Once Seoul City government had appointed the Nanjido region as the Municipal Solid Waste Landfill site in 1978, the existing farming residents of the pre-landfill era moved to other regions, while garbage collectors who had been working in other regions of Seoul moved into this area and formed a new settlement of 'outsiders'.[1]

Since the opening of the landfill, their residency initially began as squatter settlements in city-owned land. In the mid-1980s, however, the Seoul City government built a collective housing complex for the settlers and gave them the rights of residence (1984–1992).[2] In the beginning, Nanjido Landfill residents mostly consisted of *jae-geon-dae* (Reconstruction Workers),[3] but other people also moved into Nanjido Landfill for economic reasons during the early and mid-1980s.[4] After the mid-1980s, as garbage collecting in Nanjido Landfill proved to be profitable, the initial members often invited their relatives to join their landfill community. With philanthropic intentions, religious organisations, such as Catholic and Protestant churches, also entered the Nanjido Landfill residence to support the residents' social wellbeing by providing healthcare and infant care services that the landfill residents had trouble receiving legitimately from the government. Additionally, political dissidents, or protesters against the military dictatorship of the 1980s, occasionally concealed themselves in Nanjido Landfill, living as garbage collectors for certain lengths of time. The garbage collectors and dissidents in Nanjido Landfill both found the city's last resort outside the boundaries of the urban space for economic or political reasons. This spatial identity defined a substantial part of the Nanjido Landfill residence's position in the context of urban ecology.

Chapter 3 examines the characteristics of Nanjido Landfill as a human habitat by illuminating the landfill residents' housing and garbage-collecting work patterns, thereby construing the inhabited landfill of Nanjido as a border zone between environmental and social precarity[5] and defining the 'disruptive' social status of outsiders. The first part on the Landfill housing concerns the individuals' self-help management of their own living environment: self-build houses and other structures made by using found objects and altering the form as necessary. Regarding the precarity of their residences, the Nanjido Landfill town straddled the border between legality and illegality, which eventually placed the landfill habitat in a state of ambiguity.[6] Although Seoul City had provided structures akin to Quonset huts as collective housing complexes for the landfill residents, the City did not fully sanction these structures, only allowing occupancy before it announced development plans for the land. This transition from squatting to semi-legal residence placed the landfill residents in the ambiguous position of citizen-subject[7]; although, as registered residents, they were legitimately eligible to receive welfare services, there were few infrastructural service facilities within the boundary of the landfill space. Additionally, garbage collecting, which was the Nanjido habitants' primary source of work and economic subsistence, suggests socio-economic and symbolic meaning. While garbage collecting is a form of trade based on recycling objects, the essence of this job is rooted in the act of scavenging—a practice of re-ordering (recycling) through dis-ordering (scavenging). For the garbage collectors, their work was an attempt to assimilate themselves into the existing economic systems (through recycling) by resisting the social norm of dichotomous categorisation (through scavenging). Examining the practice of garbage collecting will enable us to elicit the idea that constant disarray or disruption operates behind the act of making order. Lastly, I examine how the public conceived and perceived the characteristics of the inhabited landfill of Nanjido and created imaginaries of the landfill region, which suggest the socio-ecological meaning of the Nanjido Landfill residence.

The borderline characteristics of the Nanjido Landfill settlers' residence (in-between legality and illegality) and their work (in-between recycling and scavenging) exposed the landfill habitants to permanent precarity. Considering that the landfill habitants were subjected to a sovereign power (i.e. Seoul City), yet excluded from the full protection of civil rights—particularly the right to welfare services—the landfill inhabitants' position is an example of inclusive exclusion, or the zone of exception.[8] Viewing the inhabited landfill of Nanjido from this perspective will expand the concept of landfill from a space seized by a particular political situation to a prevalent, but less visible, socio-economic space within a society.

The precarious conditions of Nanjido Landfill as a human habitat explore not only the issue of body-security-space,[9] but also the relational matter between body-security-capitalism.[10] First, the Nanjido Landfill residence demonstrates how a certain socio-economic structure divides the socio-economy within the urban space; second, the environmental and social precarity in the Nanjido Landfill region essentially emerged from the relationships amongst the body, security and capitalism, since inhabiting the landfill and performing its waste-related economic activities depended on shifting capitalist systems. Meanwhile, under modern principles, the landfill, as a site of waste is segregated from the alleged normal residences in the city. Therefore, inhabiting the landfill itself is akin to permeating a prohibited zone, which violates the existing norm of the modern city. Above all, the borderline characteristics of Nanjido Landfill's housing situation exposed the habitants to constant precarity, while retaining the potential to threaten the norm of the modern city. In this sense, this chapter examines the significance of Nanjido Landfill's residence and the habitants' work of waste collection in the context of social ecology. This analysis argues that Nanjido Landfill was a space that society feared for its disruptive potential and, therefore, endeavoured to make materially invisible and psychologically forgotten within the urban space and its historical record.

Housing in Nanjido Landfill

Self-help housing[11] (1978–1984)

The early populations of Nanjido Landfill settled close to the landfill site so that they could work in the landfill conveniently. Since the north of the landfill site was the only available level ground, on which they could establish housing structures, they eventually erected their residence between the landfill and a branch of the Han River called the Nanjicheon stream, which persistently flooded every summer (Figure 3.1). The poorly installed sewerage systems in the landfill rapidly polluted the Nanjicheon stream, and the contaminated water overflowed during the monsoon season, deteriorating the residents' housing. Since the mid-1980s, as mixed domestic waste, including food, quickly accumulated in Nanjido Landfill, the leachate leaking out of the landfill polluted the Nanjicheon stream more severely, further contaminating the adjacent landfill residence.

The early settlers, who started living in the Nanjido Landfill area before 1984, built houses by themselves with scrapped materials found in the garbage dump. Without a consolidated groundwork, they set up wooden pillars on the ground and then either tied them to nearby trees or set up props outside the wall to support the structure.[12] For roofing, they used waterproof materials such as vinyl and tarpaulin.

In 1984, most Nanjido Landfill residents moved into the collective housing complex provided by Seoul City. Since then, the city and town governments only permitted the collective housing complex and registered independent structures. Despite the illegitimacy of their residency, however, several independent huts remained in the landfill site. First, those who preferred an independent life outside the community continued to live in their landfill huts. Second, when the garbage mounds became too high for older people to ascend for daily work, they built their huts on top of the landfill mound and stayed there.[13] Third, garbage collectors commonly built huts as indispensable structures (e.g. shelters, lounges, dining places or public toilets) for the duration of their work in the landfill, and fourth, garbage collectors carried foldable tents

Figure 3.1 Self-built houses of the Nanjido Landfill's early settlers. They were located in-between the landfill and the Nanjicheon stream. Early 1980s. © Rev. Kyung-Whan Chang

as temporary storage for their collected items. Excluding foldable tents, there were about 200 unregistered temporary structures in the landfill as of 1987. The illegal housing patrol occasionally destroyed these huts, but the residents in the landfill site or users of such temporary structures rebuilt them repeatedly, so they remained until the late 1990s. The kitchen and dining place had window-like open walls, which were supposedly for lighting and ventilation. Consisting of a pair of cubes for a pit, the public toilet was the simplest structure. Meanwhile, each garbage collector's movable tent, which they used to store their personal bags of collected recyclables, was simply made of four poles and a roof without walls. The garbage collectors carried the tents with them, pitching and dismantling them as they moved around the landfill. They did not only use them to protect their objects but also to mark their ownership of collected assets.

There were no rules or standards for the housing structures of Nanjido Landfill, especially for individual housing, so the habitants built their shelters as they wished, using materials at hand. While the most prevalent form was a one-room structure constructed with four pillars and a pointed or flat roof, some habitants built unusual structures by re-using dilapidated tents and surrounding them with supplementary structures. As the scrapped material increased and the building technique improved, they later applied self-building crafts to the collective housing complex.

Collective housing complex (1984–2001)

The Nanjido Landfill population's residency generally refers to the complex of prefabricated collective housing buildings, similar to Quonsets, located in 482 Sangam-dong (Figure 3.2). In 1984, when a flood struck the Landfill residential area heavily and destroyed the self-build houses, Seoul City provided the community with collective housing near the Nanjicheon stream, which drastically changed the landfill habitants' living conditions both spatially and socially, particularly in legal terms. It was unusual that the City built the housing complex for the garbage collectors, since it was not intended to control the population—those the society's norms regarded as inappropriate—as in the previous decades; instead, it legally allowed them to live in city-owned land temporarily, if not permanently.

The motivation behind the City's decision to provide the illegal occupants with the right of residence is not clear. According to Yong-soo Kim, a Seoul City official, after the flood demolished the Nanjido Landfill habitants' self-build houses, Cardinal Stephen Kim Sou-hwan suggested that the City provide them with basic living conditions, and his remarks influenced the City's decision to build the collective housing complex and officially register its habitants as Sangam-dong town residents. By chance, the City's provision of housing for the Nanjido Landfill population along with the right of residence was more likely the works of philanthropic initiatives rather than the government's urban planning for the urban poor.[14]

Figure 3.2 Collective housing complex made of Quonset-like huts (1984–1993 [2001]). The Complex consisted of 40 buildings in total. Each building housed 24 households with an average of 4–5 family members. c. late 1980s–early 1990s. Courtesy of Seoul City

In the collective housing complex, approximately 1,000 families lived under one address.[15] Since the flood of 1984 destroyed the self-build houses of the Nanjido Landfill residence, the City spent a budget of 1,350,000,000 KRW (1,350,000 USD) to set up 40 pre-fab homes modelled on Quonsets for the landfill habitants. One building structure was designed to hold 24 families, and an average of 4 m² was allotted per family of four to six members. As a result, the total population living in the complex reached approximately 4,000 to 4,500. Each unit was equipped with a functioning water supply and sewerage system. However, since they lacked bathrooms, the families took showers in the tiny kitchen space in each unit and shared paid public toilets (1,000 KRW [1 USD] per month for maintenance) until 1989 when the City built public shower facilities and renovated existing public toilets.[16]

Throughout the collective housing complex of Nanjido, individual house-holders notably transformed the exterior and interior of each unit as needed. To complement the cramped interior, they extended the sides of the building or added extra structures like staircases, so that they could make use of the upper part of the building. On both ends of the entire complex, they attached additional structures mostly for commercial purposes, such as hair salons, clothing stores, bicycle sales and repair shops, briquette stores and so forth. Extra structures made the passageways between Quonset buildings more cramped. Moreover, due to the confined interior spaces, people often placed household goods outdoors and performed domestic tasks, such as washing clothes or dishes, outside the unit, which made the alleys even narrower. The extensions on the buildings' exteriors made the 40 buildings look as if they were connected.

Meanwhile, the interior of the units had fewer options for change other than dividing one space into two separate rooms: one for parents and the other for children. According to a Seoul City official, some households with more financial means—who remained in the Nanjido Landfill region for work or for the legal battle against the City for compensation—renovated the interior of their units far more elaborately. Some families, for example, illegally purchased the neighbouring unit and combined the two units into one, while others installed a flush lavatory inside their unit for personal use.[17]

The Nanjido Landfill residents could freely transform the housing units because they were outside the legal boundary of the land and architecture laws of the City. In particular, the extension of the collective housing building, which invaded public space, was an indisputable violation of law, but the public authority left the landfill habitants' unregistered occupation of the City-owned land in the Nanjido Landfill unsupervised. Meanwhile, although the individual sale of housing units was not completely controlled, the City strictly prohibited sales of the collective housing units to prevent real-estate speculation, particularly after it had announced plans for regional redevelopment. These inconsistencies demonstrate that the Nanjido Landfill region occupied a border zone, an in-between space straddling the divide between legal and illegal activity; they were outside the boundaries of the order of urban space as long as the landfill residency did not affect the lives of that space. Once the landfill residence became a part of formal urban planning, however, the laws of society enforced restrictions on the habitants' lives and use of space.

In their self-building and housing transformation practices, the Nanjido Landfill residents' active re-use of materials found in the landfill is notable. The emancipatory potential of the landfill habitants' self-building practice lay in the transformation of material, or the indeterminacy of the material itself: the inherent liberating force that transforms anything into something else. It is emancipatory in that much of the modern environment is blank and unresponsive, depriving the subject of an awareness of his/her surroundings and his/her relation to it; therefore, shaping his/her personal environment toward desired ends can break the vicious circle of perceptual and sensory deprivation.[18] The self-help approach to building is akin to the concept of bricolage—the term that anthropologist Claude Lévi-Strauss used to describe patterns of mythological thought that do not take steps to meet preformulated goals, but use or re-use available sources to make things.[19] While discussing schizophrenic desiring-production, Deleuze and Guattari invoke Lévi-Strauss' research and refer to bricolage as a schizoanalytic, transgressive mode of production. Inasmuch as it works with what is at hand instead of starting from scratch, we can view schizoanalysis in line with the concept of bricolage.[20] The potential of a bricolage-esque or schizoanalytic approach to production ultimately lies in the indeterminate openness of the purpose, whether regarding the materials or politics—the politics are only possible when the conclusion is not determined and open discussion is allowed.[21] The material alteration in the self-building practices of the Nanjido Landfill residence, in this vein, expands the discussion on the uncertain attitude toward materials and the environment to political dimensions. In socio-economic or political predicaments, building one's own environment becomes the subject of social ecology with regard to the environmental ecology as in the Nanjido Landfill residence, where making one's own living environment was the last resort for the habitants' socio-economic or political survival.

Adequate and sustainable housing

Adequate housing

The conditions for self-build housing are not unlike those of 'adequate housing' when pursued as a human right established by the United Nations in the Covenant on Economic, Social and Cultural Rights (ESCR). The Covenant emphasises that any structure, including self-build structures, must be protected from exposure to any environmental pathogens or other health-endangering conditions.[22] The United Nations Special Rapporteur included the right to adequate housing as part of the right to an adequate standard of living as follows: 'The right of every woman, man, youth and child to gain and sustain a secure home and community, in which to live in peace and dignity'. The rapporteur also asserts that all of these rights are interrelated; therefore, we can only fully realise the right to adequate housing when it is applied along with the rights to other conditions, including infrastructures (water, sewage disposal, electricity, home waste collection, transportation), healthcare (multipurpose community clinics, preventive medicine, dental and psychological assistance), welfare (kindergartens and primary schools, day-care centres, public libraries, artistic activities, sports and recreational facilities) and security (security of the person, security of the home and protection against inhuman and degrading treatment).[23]

In the Nanjido Landfill residence, these four conditions are all related to one another, too. The lack of infrastructures, such as protection against natural disasters (e.g. floods and fire), and the lack of safety equipment to protect garbage collectors from accidents in the landfill are directly linked to the inadequacy of the region's healthcare services. Likewise, barriers preventing full access to educational services are related to the social and psychological system that protects the landfill population from degrading treatment. Though I examine the conditions of adequate housing in Nanjido Landfill, here the infrastructures of adequate housing, healthcare and welfare services are all interrelated, and this interrelationship is ultimately a matter of security.

- Infrastructures

In the Nanjido Landfill residence, infrastructures for adequate housing, such as the water supply, sewerage system and electricity, had been installed in succession since the garbage collectors' initial settlement. Water pipelines had been set up throughout the region but the inconsistent water pressure often led habitants to make use of underground water pumps that they had installed in public spaces. The inadequacy of the sewage disposal system largely contributed to the pollution of the Nanjicheon stream. As electric lines had only been installed when the collective housing complex was built, fire accidents caused by candle misuse endangered the residents' lives during the early 1980s.[24] Although the collective housing complex provided electricity, it was not sufficiently maintained during the subsequent 10-odd years, so it caused short circuits and fires throughout the 1990s, too. In addition to the infrastructures for adequate housing, the Nanjido region required measures to protect the housing from natural disasters: floods and landfill fires.[25] First, the Nanjido region, originally a wetland by the Han River, was historically notorious for recurring floods. However, without any alternatives, such as a levy or renovated sewerage system, the same situation persisted throughout the landfill period, and the residences adjacent to the stream would be flooded almost every monsoon season. Second, as the landfill grew larger, the methane gas produced from the mixed bio-chemical operations also caused occasional fires in the landfill. General firefighting methods could seldom extinguish these fires, so bulldozer drivers would risk their lives by hurtling into the blazing flames to cut off the inflamed block of landfill mass.[26] A large amount of gas and smoke from landfill fires deteriorated the residential and work environment of the landfill area[27] (Figure 3.3).

Figure 3.3 Fire in Nanjido Landfill. c. 1990s. Courtesy of Seoul City

Moreover, since Nanjido Landfill was not designed as a work site, there were no safety facilities for the garbage collectors. There were frequent accidents endangering the lives of the garbage collectors in the landfill work site; trucks, bulldozers, or large stones plummeting down from landfill mounds struck some, while piles of waste completely buried others. As they were usually exposed to dust and other polluted materials, a number of residents suffered from respiratory and dermatological diseases as well as contracted food poisoning from the intake of toxic foods obtained in the landfill.[28] Since they threatened the security of individual lives in the landfill, the precarious natural environment and work conditions directly coincided with the matter of healthcare services.

• Healthcare and welfare

There was a stark difference between the housing situation on Nanjido Landfill and that of other regions due to the insufficiency or absence of healthcare and welfare services despite the residents' eligibility for such aid. First, as registered residents, the Landfill residents living in the collective housing complex were, in principle, eligible for healthcare services. However, they had difficulty receiving full access to the services due to either economic reasons or the lack of service facilities within their residential area—there were no public or private healthcare facilities, such as a hospital, clinic or pharmacy, within the Landfill region. Social prejudice against the landfill population was another obstacle to their access to healthcare services outside the landfill region, implying a lack of protection against inhuman and degrading treatment. The medical treatment in the Nanjido Landfill residence relied on the charitable services of individual doctors or experts from religious organisations. Doctors related to Catholic or Protestant churches visited the Landfill region on a weekly, bi-weekly or monthly basis.[29] In the case of serious illness or fatal injury by accidents, the habitants went to the nearest Catholic hospitals located outside Nanjido with the help of nuns who were living in the landfill.[30] It was not until 1993, when the Nanjido Landfill stabilisation and regeneration plan was announced, that the town office provided the Landfill residents with free medical check-up services.[31]

Second, the Nanjido Landfill collective housing residents were legitimately eligible for both public and private welfare services, particularly educational services. However, they had difficulty obtaining full access to those services as well because of the lack of institutions within their residential area and the tacit discrimination against the landfill population, which discouraged them from receiving the services to which they were entitled. The students of the Landfill area went to schools in neighbouring regions (e.g. North Sangam-dong or Susaek located north of Sangam-dong), but they sensed an invisible border between them and students from other regions when they went to school.[32] After-school programmes for elementary or junior/high school students run by voluntary college students partly supplemented their insufficient education.[33] Despite the absolute necessity for mothers to work in the landfill, nursery services were scarce. Although religious organisations compensated for the dearth of childcare with nurseries, they only provided these services for limited hours, so working mothers often brought their babies to the landfill work site, which was particularly dangerous as the infants would eat inedible materials within reach.[34] Likewise, weak welfare services made the landfill habitants physically and psychologically vulnerable, and this vulnerability required immediate healthcare services and fundamental security systems to protect them against inhumane and degrading treatment.

The conditions that self-help housing must meet to be deemed adequate housing include *affordability*, *liveability*, *security* and *sustainability*.[35] In the Nanjido Landfill habitat, the housing was affordable in the sense that early self-build structures were constructed on squatted land and the City had provided the collective housing complex for free. The inhabitants also continued to alter the structures to make their housing conditions more habitable. However, the security system was poor, and there was neither a legal nor a financial guarantee of sustainability. Moreover, the lack of sustainability challenged the affordability, liveability and security of the home. Maintaining the right to reside in the collective housing—housing security based on legal protection—was essential to sustain the adequate conditions of their housing, especially since the City's development plan intended to disintegrate the landfill habitat at any time.

Sustainable housing

- Legal security of housing

The Seoul City government and town office strictly regulated the residents moving in and out of the Nanjido Landfill collective housing complex because it was the City's property. The City had only permitted residents who were working on behalf of garbage-related businesses for a limited time before it carried out its redevelopment of the site. Thus, the right of residence to the collective housing was of absolute importance for most Landfill residents, as it was the only warranty through which they could receive financial compensation after government policies forced them to vacate the Complex. The following episodes demonstrate how desperately the residents endeavoured to retain the right of residence and how strictly the government controlled the illegal occupation of the Complex.

> You can move out of 482 Sangam-dong (the Nanjido Landfill collective housing complex) but you cannot move in. When I started working in Nanjido Landfill, I had to change my residence registration. A female public servant at the town office helped me move in, conceiving that I, as a nun, would not cause any commotion regarding the right of

residence. When I visited her again to confirm my registration, all of the other officers strongly opposed her. Against all odds, she made it possible for me to reside in the Complex. That was the last time I saw her in the town office.[36]

An old woman was living next door with her two sons. When her younger son became a soldier in the professional army, he needed to change his residence registration to the army base. When the woman received the notice letter, she visited our unit for help with filling out the document because she was illiterate. However, our apprentice mistakenly wrote the names of all three members of the family as if the whole family would move out. To re-register them, I even had to meet with Seoul City officials to make an appeal, and fortunately enough, she got her housing unit back.[37]

On an individual level, obtaining and retaining the right of residence to the collective housing complex by becoming a registered citizen meant not only possessing property but also acquiring individual security. Only once they became registered citizens could they enjoy secure protection against inhumane or degrading treatment, and access to healthcare and welfare services. Administratively, many public officials opposed the provision of residence rights to the Landfill population from the outset, as they essentially regarded the Landfill habitants as squatters. All the while, they were concerned about the compensation the City would have to pay the occupants once the City evicted and relocated them to redevelop the site. In reality, however, the compensation provided Seoul City with the grounds to demolish the inhabited space of Nanjido Landfill, and the justification to expunge the so-called inappropriate realm from the urban space of Seoul.

- Relocation: economic sustainability

Seoul City closed Nanjido Landfill in 1992 and initiated the landfill's stabilisation and regeneration plan, which accompanied the relocation of the collective housing residents. The City suggested it provide the residents with priority purchasing or residing rights for leased apartments. They had two options: the existing residents could purchase a 69–82 m² apartment with a national loan of 12,000,000–14,000,000 KRW (12,000–14,000 USD) payable in 19 years with a 1 year grace period, or they could choose to lease 46–52 m² apartments with various rents. However, most people claimed that they could not afford those apartments and wanted to stay in the Nanjido region, which delayed the demolition of the collective housing complex until August 1996; by May 1996, 355 families accepted priority rights to purchase or move into leased apartments.[38]

During the extended negotiations between the residents and the City (1996–2001), the City appointed the Sangam-dong area as the site for the

World Cup games main stadium in 1998. The plan to transform the Nanjido Landfill region into a park and to redevelop the overall Sangam-dong area followed, driving the City to accelerate the negotiations to relocate the Landfill residents without conflict. In 2000–2001, the City had to finalise negotiations so that the government could complete the park's construction before the opening of the 2002 FIFA World Cup Korea/Japan. Finally, in 2001, after nine meetings between the resident representatives and the Seoul City mayor or the second vice-mayor, both sides reached an agreement on condition that the City provides the residents with priority purchasing rights for the new Sangam-dong new town apartments and waives the maintenance fees necessary prior to settling down in their new location.[39] This not only implies that the residents requested more compensation but also demonstrates that they wanted to remain in the region of their socio-economic base—many collective housing residents were working in the landfill or neighbouring regions, particularly involved in waste-related businesses. In practice, however, as they could not afford the high cost of the new Sangam-dong apartments, most people sold their priority rights and settled in affordable houses in other regions. When possible, they moved into neighbouring towns to maintain their jobs. Nanjido residents who had moved into the new Sangam-dong apartments were all accommodated in building No. 2.[40]

The Nanjido redevelopment and relocation case was significantly different from others in Seoul, which often resulted in fierce conflicts and court battles. This was partly because the upcoming national/municipal sporting event drove the City to resolve the issue of relocation within a limited time without violent conflicts. Therefore, the City regards the Nanjido Landfill population's relocation as a 'settled' case. However, for that very reason, the area's history as the Landfill, including the environmental and social ecologies of its residence and their relation to the urban fabric of Seoul, is no longer a subject of concern in both urban and environmental discourses. Moreover, because former garbage collectors do not tend to reveal their past careers, the identity of the site as an inhabited landfill has been rapidly untold, buried and annihilated—this rapid erasure implies that the socio-psychological system of degradation perpetrated against garbage collectors still operates. In regional redevelopment, financial compensation is often regarded as the best solution for both the developer and the existing habitants. For the developer, it justifies issues related to forced relocation; for the habitants, as real-estate prices rise, it could be an opportunity to earn additional profit, which, in most cases, provides the recipients with affordable, liveable, secure and sustainable housing. However, for the urban poor, we must seriously consider the means for sustainable housing, or sources of consistent income, as well as the stipulated financial compensation for affordable and liveable housing.

Once people had begun working in Nanjido Landfill, most would not move out mainly because they had few job options outside the landfill region—from 1977 to 1987, only 20% of the Nanjido Landfill habitants

changed their profession. More than 95% of the Nanjido Landfill population were garbage collectors or served waste-related industries. Therefore, in
1993, many residents protested against the government's plan to redevelop
the landfill site, expressing their desire to remain in the area and demand-
ing alternative job opportunities to sustain their livelihood.[41] According to
a survey conducted by the Nanjido Moving Plan Committee (an autonomous organisation established by the Nanjido residents), 43.4% of families wanted to remain in the Nanjido region and receive the right to the
land so that they could build their own houses in a familiar living environment and continue their work. As the closure of the landfill drew near in
1992, the Landfill residents officially requested that the government provide
work opportunities in the new Sudogwon Landfill (Capital Area Landfill
outside Seoul) and hire 150 people. However, the City rejected their proposal because the new landfill was a sanitary landfill, one that only accepts
non-recyclable final waste after recyclable items had been sorted out before-
hand.[42] Since Nanjido Landfill closed, 34.4% of the former landfill garbage collectors became temporary workers, e.g. construction workers and
housemaids, and 22.6% remained unemployed.[43] Changes of occupation
from garbage collector to construction worker or housemaid demonstrate
the increased employment instability; while landfill garbage collectors were
independent labourers, working as they pleased without external restrictions, construction labourers and housemaids are temporary labourers hired
by corporate entities, susceptible to a competitive job market.

The Nanjido Landfill closure occurred in tandem with the establishment of a new economic system in the 1990s (reaching its peak after the
1997 financial crisis), which accelerated the Nanjido Landfill residents' loss
of an income source. The new system restructured waste management and
the job market in the field of recycling. In collaboration with large private
corporations, Seoul City took over the recycling business and operated it
based on mechanical systems, which produced massive unemployment—the
newly unemployed were deemed improper, and thus, incapable of secur-
ing a socio-economic position for systematic reproduction.[44] As Zygmunt
Bauman states, it is an inevitable outcome of modernisation, a side effect of
economic progress, which cannot proceed without devaluing the previously
effective mode of 'making a living' and without subsequently depriving their
practitioners of their livelihood.[45] Negotiations on Nanjido Landfill's closure
and the landfill residents' relocation had placed overt emphasis on the relatively reasonable compensation, without bringing attention to the residents'
forced job changes or the production of unemployment, which ultimately
deprived them of sustainable living conditions. In addition, the financial
compensation overshadowed other issues, including the dissolution of the
self-help community and the disintegration of the habitants' emotional
attachment to the site. As a result, although the Nanjido Landfill housing, an architectural frame of habitation, showed the transformative and,
thus, disruptive potential of the self-help environment, its socio-ecological

precarity prohibited the Landfill community's self-organisation, which hinges on sustainable adequate housing.

Garbage collecting in Nanjido Landfill

To secure sustainable housing, the Nanjido Landfill residents relied on their major source of income, garbage collecting practised in the mixed waste dump of the unsanitary landfill. Economically, as the amount of consumption drastically increased throughout the 1980s, garbage collecting became more profitable than ever before—the garbage collecting business began to decline around the late 1980s and early 1990s as the waste industry opened to the global market and imported cheaper waste than the local varieties.

The Nanjido Landfill individuals worked independently and earned as much as they worked. Since garbage collecting was not a legitimate form of business, the workers avoided income tax, which contributed to their overall earnings. Rev. Chang stated, 'In the 1980s, Nanjido Landfill was the land of opportunity for diligent people and paradise for the lazy as they could earn enough money to live a day-to-day existence by working whenever they needed'.[46] The Nanjido Landfill garbage collectors' monthly incomes varied depending on the individuals' existing assets and work skills. The average monthly income of the garbage collectors was approximately 200,000–400,000 KRW (200–400 USD) per month, which corresponded to the salary of the lower middle class during the early and mid-1980s.[47] For this reason, the population of landfill workers increased from 550 people in 1980 to 2,000 in 1984.[48] According to the former landfill residents, many garbage collectors, mostly those who had pre-existing assets, accumulated enough wealth to purchase houses or buildings outside the Nanjido region during the mid- to late 1980s.[49]

As for the work systems in Nanjido Landfill, there were two types of garbage collectors: the 'front-earner' (*ahp-beol-e*) and the 'rear-earner' (*duit-beol-e*). The front-earners had a right to the waste of one or more of the trucks from 17 towns throughout Seoul (the total 40 front-earners had the rights to approximately 700 trucks). They sometimes traded the right to a garbage truck for 2,000,000–3,000,000 KRW (2,000–3,000 USD). The strict rule was that only after the front-earners gathered the recyclable items could the rear-earners access the rest of the waste. The front-earners started working early in the morning when garbage trucks arrived and unloaded the waste at dawn, while the rear-earners began their morning shift around 7 am and finished around 1 pm—they could continue working in the afternoon as they wanted (Figure 3.4). While the rear-earners worked, City-operated bulldozers drove back and forth to compile scattered garbage to one area so that the rear-earners could collect items more easily. To prevent accidents, a city official would monitor each bulldozer's movement and whistle when it approached the garbage collectors so that they could step aside.

Figure 3.4 Garbage collectors working in the landfill. c. late 1980s–early 1990s.
    Courtesy of Seoul City

Nonetheless, there were many casualties from bulldozers running over people. As such, there was a clear differentiation in income and work conditions between the two groups of garbage collectors; front-earners had more
opportunities to ascend the ladder of economic class, whereas rear-earners
would remain in the low-income class for a longer period, while facing the
risks of a more dangerous work environment without safety equipment.

    Regarding the nature of their business, garbage collecting is a form of
recycling; this aspect of their work enabled the habitants to sustain their
lifestyles. Seen from the perspective of human behaviour and its relationship to the material environment, garbage collecting is essentially an act of
scavenging that at once establishes the garbage collectors' degraded social
position and threatens society's existing norms by disrupting the order of
categorisation.

Recycling: assimilation

There are four basic methods of garbage disposal: dumping (sanitary or
unsanitary landfill), burning (incinerating), turning waste into something
useful (recycling) and minimising the volume of material goods (future garbage).[50] Garbage collectors engaged in turning waste into something useful,
or recycling, in Nanjido Landfill. There were two objectives for garbage collecting in Nanjido Landfill: one for personal use and the other for making
a profit. For personal use, the landfill residents built their houses and other
structures for common use by using found materials from the landfill, such
as wood, metal, tarpaulin, vinyl, fabric and so forth. They re-used all kinds

of necessities from electronic appliances to small household goods, and they either upcycled defected items or reclaimed the waste by reinterpreting the original purpose of the goods. In Nanjido Landfill, recyclable items were categorised by product (e.g. repairable electronic items, mechanical parts, etc.) and material type (e.g. metal, plastic, fabric, wood and paper) (Figure 3.5).

Figure 3.5 Recyclable papers packed into specified dimensions. c. late 1980s–early 1990s. Courtesy of Seoul City

Amongst all the raw materials, the garbage collectors mostly preferred metals, particularly brassware, steel and aluminium because the price fetched from recyclable materials was normally based on weight. The garbage collectors made an extra-large bag specially designed for plastics to address the enormous quantity of plastic items found in the landfill. After sorting out the recyclables, they packed them into specified dimensions before trading them. After taking these measures, the garbage collectors sold all recyclable products or materials to second-hand shoppers, who then resold them to factories that processed the refurbishable items and materials.[51] Second-hand shoppers came to the Landfill regularly, on average every 10 days, to purchase the recyclables, paying by cash.

The garbage collectors recycled food mainly for their own consumption but, in certain cases, they resold raw food to the market.

> The days when the garbage trucks arrived from the American Army base were days of festivity. We procured a myriad of unopened snacks and cookies from them. One day we even gathered entire packs of chicken and sold them to the *Moraeanae* market, a traditional market in the neighbourhood.[52]

During the mid- and late 1980s, especially after the 1986 Asian Games and 1988 Seoul Olympic Games, overall consumption and the amount of food refuse rapidly increased.

> At first, it was difficult for me to eat food from the garbage dump. But, later, nothing was more delicious than eating fresh fruit while working in the landfill. There were enormous amounts of eateries in the landfill, especially in the winter; after lunar New Year's Day, we used to get rice cakes, pan-fried pies and meats. During the 1988 Seoul Olympic Games, we found whole packs of unopened high-quality foods from the athletes' camps.[53]

The garbage collectors recycled food in a somewhat different way from how they recycled manufactured products. Sometimes they gathered and consumed leftovers, but in most cases, they collected rejected raw ingredients still intact. In either case, food recycling is essentially commensurate with *gleaning*, or *consuming the over-production of society*, rather than with reclaiming refuse whose value had expired. Food recycling, in this respect, harbours a subversive signal that opposes the post-industrial capitalist economy represented by over-production and over-consumption. The Nanjido Landfill habitants' recycling methods, re-using or gleaning, suggested a precedent for alternative consumption movements, such as freeganism.[54] Of course, the recycling practices of Nanjido Landfill did not foster anti-consumerist or anti-capitalist ideological intent. Rather, the garbage collectors' primary purpose was to partake in the existing economy through a recycling business that the society required. In this way, they were assimilating themselves to the socio-economic norms of the city.

Scavenging: disruption

While the ragpickers of the past picked up refused materials in the public spaces of Seoul, the Nanjido garbage collectors worked in the separate space of Nanjido, but both their work salvaged the value of abandoned materials. In the early twentieth century, Walter Benjamin, while reflecting on Baudelaire as the writer of modern life, found an analogy between the poet and the ragpicker in their mutual collection of useful items in refuse (in the modern French context, ragpicking means rummaging through refuse in the streets to collect materials).

> The poets find the refuse of society on their streets and derive their heroic subject from this very refuse […] This new type is permeated by the features of the ragpicker, who made frequent appearances in Baudelaire's work.[55]

Benjamin points out that ragpickers and poets are both concerned with refuse, working while other citizens are sleeping. He also explores the similar gestures and walking style found in both groups: 'This is the gait of the poet who roams the city in search of rhyme-booty; it is also the gait of the ragpicker who is obliged to come to a halt every few moments to gather up the refuse he encounters'.[56] Baudelaire himself described the ragpicking in prose:

> Here we have a man whose job is to gather the day's refuse in the capital. Everything that the big city has thrown away […] He collates the annals of intemperance, the capharnaum of waste. He sorts things out and selects judiciously; he collects, like a miser guarding a treasure, refuse which will assume the shape of useful or gratifying objects between the jaws of the goddess of Industry.[57]

Baudelaire captures the very essence of ragpicking: collecting and categorising potentially valuable objects. More precisely, it is the re-collection and re-categorisation of objects that the norms of the urban society have already pre-ordered. By comparing the ragpicker to the poet's writing practice, Benjamin signifies that both harbour the creative power rooted in the act's own subversive potential. When examining ragpicking in public spaces and garbage collecting in the landfill more closely, we can differentiate the two in terms of the degree of disruption. Since the landfill is a topographically separated or bordered site of refuse—rife with objects that have been determined inappropriate and, therefore, excluded from the space of the appropriate—the garbage dump must be left intact to prevent it from blending with the appropriate objects of the urban space. Unlike ragpicking in pub-lic streets, garbage collecting in the landfill meant that someone actively entered an inappropriate zone to rummage through objects that have been pre-categorised as inappropriate.

In the twenty-first century neoliberal economy, Zygmunt Bauman draws on Mary Douglas's idea of 'dirt' in his interpretation of waste as either something that has lost its use value or someone who lacks the ability to become a proper producer and consumer, therefore, remaining as surplus value. In this line of thought, he elucidated the meaning of waste disposal in modern society:

> Waste is the dark, shameful secret of all production (alongside the secu-rity service [...] aimed at staving off the return of the repressed [...]). Hence, the waste-disposal industry is one branch of modern production that will never work itself out of its job. Modern survival—the survival of the modern form of life—depends on the dexterity and proficiency of garbage removal.[58]

As Bauman states, garbage in modern capitalism is something that needs to be repressed, and its return is so threatening to modern survival that it reinforces the security industry. The act of scavenging, the disruption of repressed waste, holds a potential threat to the socio-economic norm of the modern city. In Egypt, scavengers are known as *zabaline*,[59] predomi-nantly consisting of Coptic Christians, and in Mexico, scavengers are called *pepenadores*, who are mostly unionised and even powerful.[60] As in these examples, the disruptive power of scavengers stems from their identity as an anomaly and their practices' characteristic essence of defiance against the existing social norm. Ultimately, the subversive potential of scavenging lies in the way the scavengers cross the border between the two divided spaces. This gesture, which challenges the modern society's strict categories or divi-sions and endangers its norm, blurs the border between the appropriate and the inappropriate.

Imaginaries of Nanjido Landfill

Throughout the 1980s and 1990s, newspapers had occasionally dealt with Nanjido Landfill and the garbage collectors' lives. The tone and rhetoric of the media at the time commonly showed two perspectives: first, the soci-ety viewed the landfill as a zone separated from other city spaces; second, landfill habitants lived socio-economic lives, unexpectedly, similar to those in other parts of the city. After the landfill's closure in 1992, news on the landfill residency (e.g. accidents in the landfill area) and the links between the population's relocation and the City's development plans and real-estate values became public knowledge. Nonetheless, the public perceived the Nanjido Landfill space as a realm outside the boundaries of the city. On the one hand, Nanjido Landfill was a lived space[61] where the landfill habitants led an alternative socio-economic life. On the other hand, in the imaginaries of the Landfill, it was a mythical space, neither existing in reality nor forgotten in the public's mind, thereby becoming a space of exception.[62] The public based its imaginaries of Nanjido Landfill not on the socio-economic lives of the people in the landfill but on its conceptions of garbage as dirt, or the potential threat of disease/death, which could endanger the modernised urban space.

With a conceptual framework based on the relational dynamics between the body, security and space, Joe Penny argues that extreme insecurity produces the space of exception.[63] Insecurity (of an individual's subjectivity and of sustainable living), one of the crucial concepts discussed in studies on troubled regions (e.g. the Palestinians in East Jerusalem), is often concerned with precarious geographies; for example, walls or checkpoints and systematic discrimination function as the spatial materiality that creates and controls the in/security of life. Without militant political confrontation involved on a state level, the Nanjido Landfill area did not exactly have geographical precariousness.[64] Rather, it is more appropriate to view Nanjido Landfill as an internally colonised space: geographically located within, yet perceptually excluded from, the urban space, while simultaneously answering to a system governed by the city, so that landfill habitants labour to preserve the city's socio-economic orderliness. In this case, we can apply Agamben's paradoxical notion of 'inclusive exclusion' to the landfill's position in the city. Moreover, although the exclusionary imaginaries about Nanjido Landfill produce a psychological border between the landfill and the non-landfill regions, garbage collectors' freedom to cross the border represented by the sensory attributes of refused materials (e.g. the miasmatic trait) gives the landfill the identity of a *border zone*—the uncertain in-between space dividing the inside and the outside.

Fear and threat

In the supposedly normal part of the urban space, the public perceived the Nanjido Landfill space as dirt itself, and regarded the habitants, who lived in the less adequate residential conditions, as analogous to the space of the landfill, or dirt itself. The public responded to the dirt of the landfill by either degrading and/or excluding it, while equating it with something fearful, or a threat to society.

First, as one way to degrade the landfill population, non-landfill people stigmatised the landfill habitants as pre-modern or primitive; for example, during my fieldwork in Seoul, I found that public officials often called the Nanjido residence an 'Indian Village'. When I asked former residents which part of the town the byname referred to, however, they answered that they had never heard of the name or of any landfill habitants using that expression.

I've never heard of the 'Indian Village' [...]. It sounds as if it refers to a primitive, uncivilised living environment in the jungle. [...] In Hong Kong, there are people who live on board,[65] living in houses built on ships. Other people would call them by a particular degrading name, but they never call themselves by that name.[66]

According to this testimony, only non-landfill residents used the byname 'Indian Village' to indicate the Nanjido Landfill community. Here, Indian refers to the Native Americans who maintain their tribal community and traditional lifestyles, separate from the modernised American culture. This name reflects the non-landfill population's preconceptions of Nanjido Landfill, its habitants and their work of garbage collecting; in essence, non-landfill people regard garbage as dirt and the landfill's unsanitary living and working environment as pre-modern and uncivilised.

Second, a conversation between an anonymous interviewee and Jade Keunhye Lim, who were both non-Nanjido residents, demonstrates how the concept of 'fear' operated in the public's view of the Landfill area, ultimately generating an exclusionary practice regarding the space. As the anonymous interviewee described the personal fears s/he experienced in the landfill region in the mid-1990s when the Landfill had officially closed, while the residence and landfill mounds remained, Lim expressed her own conception of the Landfill region in response.

A (**Anonymous**)*:* All of the buses I take from school terminated in the Nanjido area via Mangwon-dong where I was living. [...] One night, I was drunk and fell asleep on the bus before it terminated at the Landfill site.

Lim (Jade Keunhye Lim)*:* Did the bus cross the Garbage Mountains? Really?

A*:* Yes, it did. There were no street lamps. [...] There were *strange effluvia*. It was tremendously *fearful and scary*. [...]

Lim*:* I thought that people just dumped garbage in the Nanjido region and the dump had eventually expanded. Were there roads inside the landfill? [...]

A*:* There was a town, but the town was scary, too. It was a traumatic experience for me to be there that night. When I heard that the City planned to build a Park in the Nanjido Landfill site, I thought it would be a cataclysmic change.[67] (Italics added)

This conversation shows the public's typical imaginaries of Nanjido Landfill. The individual who had been to the landfill site remembered the area as fearful and scary, whereas the individual who had never been to the site did not know that it was an inhabited landfill. The anonymous interviewee's experience of fear, although intensified by the darkness of the night and the languid atmosphere of the town after the landfill's closure, demonstrates considerable truth about the prevalent imageries and imaginaries of Nanjido Landfill. The fear that s/he experienced was, first, the fear of death or disease potentially caused by unsanitary objects; second, the fear of crime that is presumed to be committed more often in underdeveloped and/or isolated areas; and third, the fear of uncertainty regarding the unidentifiable effluvia. Since the olfactory sense, unlike sight or the sense of touch, evokes dubiousness, unidentifiable miasma must have amplified his/her fear of uncertainty.

The lack of sanitation, unhealthiness, disease and death are the concepts most often mentioned in modern discourses of biopolitical power, which is characterised as an ability to live and let die.[68] Based on this logic, to create and maintain purity (of the appropriate urban space) and security (of the appropriate citizen-subject), society must insulate Others, including the sick, the mad and the criminal,[69] through a variety of regulative regimes, including constitutions, laws, policies, bureaucracy, border controls, population censuses or invented histories and traditions.[70] Since Nanjido Landfill was a legitimate part of the city, however, the nation or City could not establish any constitutions, laws or policies to separate and exclude the region from other urban spaces. Instead, imaginaries of the landfill site created a socio-psychological border based on the modern logic of dirt as something fearful or a threat to an ordered society. Likewise, the imaginaries about Nanjido Landfill operated well enough to insulate the site along with its inhabitants and their occupation.

Subversive zone

While discussing the creation of urban imaginaries of the landfill, I asserted that modern society had to be insulated from the socially inappropriate— the sick, the mad and the criminal—drawing on Derek Gregory's idea. The association of criminals with the sick and the mad as one of the social anomalies is notable here. Although the sick and the mad represent a latent state of physical and psychological abnormality, a criminal is a subject who actively causes disorder by violating laws and regulations. If the sick and the mad signify potential agents of anomaly (germs), a criminal is an active practitioner of violation and disruption. Generally, a criminal is defined as a person who endangers the security of the lives of citizens and of property. Depending on the socio-political context, however, a criminal, especially a political criminal, or dissident against a current regime, holds a subversive yet possibly creative power to challenge conventional rules.

As far as the zone of abandonment has its own power of production, preservation and even disruption, it does not merely remain an excluded space, subordinated to other city spaces; in this case, the two separate zones form a relationship of reciprocal exclusion.[71] Moreover, the ambiguity of Nanjido Landfill—which stemmed from the materials' (particularly gaseous substances) and garbage collectors' free passage across the border as well as scavenging's disruption of society's categorisation—not only made this site a *border zone*, or uncertain in-between space, but also gave it the power to threaten non-landfill spaces and the dichotomous order.

Dr Woon-soo Kim, a senior researcher at the Seoul Institute, recalls the image of Nanjido Landfill in the mid-1980s as a ghetto-like region, and compared the landfill residence to *sodo* which is generally known as an isolated zone rife with criminals in ancient Korea (the Samhan period in 3–2 centuries BC).[72,73] Kim's comparison of Nanjido Landfill with *sodo* suggests insight into the politico-spatial implications of the landfill region. *Sodo* was at once a place for sacred rituals and an area where criminals gathered. Designed for rituals, it was located on higher land, where a high priest led rituals in front of a phallic-shaped sculpture. Since *sodo* was the territory of the gods, state laws did not apply to it, so it became an area of extraterritorial jurisdiction. Therefore, once criminals had escaped into this zone, the state could no longer arrest them. These characteristics consequently made this site both sacred and profane. In this sense, *sodo* is similar to the old English word 'asylum', which has roots in the Greek *asulon* (without right of seizure) or refuge, and, in Late Middle English, means a 'place of refuge' especially for criminals.[74]

The metaphorical analogy of Nanjido Landfill to ancient *sodo* reveals two facets of the landfill site: the conceived/perceived space (from the non-landfill people's point of view) and the lived space (of the landfill habitants). For the non-landfill people, the Landfill was a dirty, and thus, fearful and threatening space that they wanted to avoid, so the public tended to annihilate its existence. For the insiders, it served as a socio-economic asylum for two main reasons. First, most of the Landfill population entered and settled in the area to seek work opportunities for their socio-economic life. Second, Nanjido Landfill felt like a resort for Reconstruction Workers with criminal records because they could get by without living under the gaze of society's discriminatory eyes. Under the dictatorship of the Fifth Republic of Korea during the 1980s, political dissidents found political asylum in the Landfill, too.[75] For example, Jae-soon Yoo, a protester against that dictatorial regime, found asylum in the landfill after the National Security Agency had severely tortured him; he then lived as a garbage collector for several years.[76] The dual, or sometimes conflicting, aspects of the region—protected yet potentially disruptive—increased the ambiguity that the ordered modern society most guarded against. Likewise, Nanjido Landfill provided a zone of socio-economic and political asylum to people in need as *sodo* of ancient Korea had. It was possible due to the region's ambiguous borderline characteristics; officially, the landfill site was managed under the rule of law, but it tacitly remained a space of abandonment or a zone of extraterritoriality.[77] Its ambiguity would position the landfill and its people's lives under an unpredictable sovereign power, and it allowed for socio-economic and political practices relatively free from government control.

In *Society Under Siege* (2002), Bauman defines the identities, roles and impacts of refugees (asylum seekers for political and economic reasons). He argues that, since they come from outside but settle in the neighbourhood, people often regard them as a source of fear for doing their jobs without consulting those whom their actions are bound to affect.[78] Outside of their refugee camp, these individuals are considered trouble, yet they are forgotten when they stay inside it. Accordingly, the primary concern for the existing society is to keep refugees inside the camp and hold them at a distance from society.[79]

As the society wanted the Landfill residents to remain within the boundary of their realm, making it possible to forget them, the Nanjido Landfill habitants' social position was analogous to that of refugees. More precisely, they were internal refugees and their territory of occupation was an internal colony created and maintained by an urban structure that had inherited the norms of modern society and been constituted by a combination of the late twentieth century consumerist society and the neoliberal economic system of the new millennium. By revealing the existential value of the Nanjido Landfill and its population, we not only recover the citizen-subjectivity of the landfill habitants, but also challenge the division-based, order-centred modern value, upon which modernised society has relied for centuries.

The Nanjido Landfill residence was exposed to diverse precarious external conditions; it lacked, for example, safeguards against natural disasters, semi-legitimate housing and adequate access to infrastructures, healthcare and welfare services. More precisely, considering that the Nanjido habitants acknowledged hunger, the need for shelter, vulnerability to injury and destruction as political issues, we can conceive that their habitational condition was, in Butler's term, in precarity. The City's financial compensation of the Nanjido Landfill collective housing residents, however, annihilated and buried the spatial history of Nanjido Landfill and the issues of precarity. In the socio-spatial sense, the financial compensation reconstructed the habitants' physical location and social position on the border zone into a legitimate socio-economic space. That is, the Nanjido Landfill habitat, in a sense, functioned as a transitional space for the habitants to transform themselves from non-citizen-subjects to economically appropriate citizen-subjects, and to move from an illegal sector to a legal one, or arguably the normal urban space.

Nanjido Landfill and its residences were a border zone of ambiguity not only with respect to environmental aspects (landfill as a form of structure in-between the natural and the artificial) but also to social aspects. First, Nanjido housing fell somewhere in-between legal and illegal habitation. Second, garbage collecting in Nanjido Landfill was both a recycling business and an act of scavenging; while recycling was their major form of economic activity within the boundary of the existing rules, scavenging disrupted the existing social rule of categorisation. Third, Nanjido Landfill became a border zone through the imaginaries about the site: a lowly and fearful area. The uncertainty of the border zone, and the free passage across the border between the appropriate (the clean) and the inappropriate (the dirty), especially the material and symbolic crossing of the effluvia, generated a threat that could disrupt the ordered urban space.[80]

As a border zone, Nanjido Landfill demonstrates how its boundaries were not impregnable but porous, allowing for free crossing between both sides. However, by refusing the idea of a reciprocal relationship—the idea that 'I am already bound to you' and 'I *am* my relation to "you"'—the overall urban body could not avoid exposure to precarity,[81] since susceptibility and vulnerability constituted both bodies at the most fundamental level. In other words, precarity is not only the problem of a fragment of the city that suffers from material instability, but also a matter of the entire urban space in which inhabitants lack an awareness of ethical responsiveness and responsibility for one another.

Notes

1 For the indigenous Sangam-dong people, who were living in North Sangam-dong, the Nanjido Landfill people were outsiders not only because they moved in from other regions, but also because the indigenous Sangam-dong community members held the shared mindset that they were different from the landfill inhabitants who made their living by collecting garbage. From the interview with Jong-tae Lee (7 May 2014), the chair of Sangam-dong community.

2 Seoul City decided to demolish the collective housing complex in the Nanjido Landfill area immediately after the landfill's closure at the end of 1992. Despite the official prohibition, a large number of people remained in this housing complex throughout the 1990s, while negotiating with the City for compensation. In 2001, the negotiation was settled and the Complex was demolished to build the World Cup Park.

3 However, according to Jae-soon Yoo, a journalist and novelist who lived in Nanjido Landfill for approximately 3 years in the mid-1980s, in Nanjido, *jae-geon-dae* referred to people with criminal records, while former garbage collectors were called *dae-won*, which simply means 'member'. On the other hand, Rev. Kyung-Whan Chang stated that, in the early 1980s, people used the term *jae-geon-dae* to refer to both the professional garbage collectors and former criminals who started working as garbage collectors in Nanjido Landfill (Interviews with Jae-soon Yoo [21 August 2014]. Rev. Kyung-Whan Chang [13 August 2014] lived in the landfill from 1980 until 2001 and took the position of the residents' representative when negotiating with the Seoul City government on the landfill residents' relocation and compensation).

4 According to a survey conducted on 5 June 1993 with the 532 families (out of 820) that remained in the closed Nanjido Landfill region, 27.1% moved to the Nanjido Landfill area immediately after the landfill opened and 26.5% moved in when they went out of business (*Hankyoreh* [19 June 1993]).

5 Judith Butler in *Precarious Life* (2004) used the noun 'precariousness' but later used 'precarity' more often than 'precariousness' as in the article titled 'Life, Vulnerability, and the Ethics of Cohabitation' (2012). Butler argues that the notion of precarity is essentially related to the ethical obligation of humans to others, including human and non-human beings. She also claims, 'Precarity only makes sense if we are able to identify bodily dependency and need, hunger and the need for shelter, vulnerability to injury and destruction, forms of social trust […] as clearly political issues'. In *Frames of War: When Is Life Grievable?* (2016 [2009]), Butler states, 'Precariousness and precarity are intersecting concepts'. In

this study on Nanjido Landfill as a habitat, I selectively use these terms; precariousness for unpredictable external situations, precarity for concerns of precariousness related to ethical obligation as in Butler's definition. See Butler, 2004; 'Life, Vulnerability, and the Ethics of Cohabitation', *The Journal of Speculative Philosophy*, Vol. 26, No. 2, Special Issue with the Society for Phenomenology and Existential Philosophy, 2012: 134–151; and *Frames of War*, London and New York: Verso, 2009: 25.

6 In a study on the self-build or self-help habitat, David Westendorff mentions, 'The largest single urban land tenure category in many developing countries is that of extra-legal land developments. These include a wide range of land development practices, from squatting and unauthorised sub-divisions to the construction on registered land of houses that have not been officially sanctioned [...]' (David Westendorff, 'Three Essays on Urban Governance and Habitat in Developing Countries', Dissertation at Cornell University, 2009: 56).

7 Verena Conley states that even if humans ultimately aspire toward a borderless world, especially in today's globalised environment, the state and borders still define them. They are not simply migrating masses, but subjects who must ask for the right to move, reside and inhabit. In this sense, she claims that an 'eco-subject' must also be a 'citizen-subject', a term coined by Étienne Balibar. See Conley, 'The Ecological Relation', 2013: 283; Étienne Balibar, *Droit de Cité*, Paris: Edition de l'Aube, 1998.

8 Regarding 'inclusive exclusion' and 'the space of exception', see Giorgio Agamben, *Homo Sacer: Sovereign Power and Bare Life*, CA: Stanford University Press, 1998; and *State of Exception*, IL: University of Chicago Press, 2005.

9 While discussing biopolitics-security-space relations, Joe Penny points out that further work is needed to tease out the relationship between biopolitics, security and capitalism, and uncover how this relationship works to produce spaces of insecurity. He continues to state that we could apply this framework to the neoliberal 'annihilation of space [...] the annihilation of the people who live in it' (Joe Penny, 'Insecure Spaces, Precarious Geographies', UCL Development Planning Unit, DPU Working Paper, No. 141, 2010: 27; Don Mitchell, 'The Annihilation of Space by Law', *Antipode*, Vol. 29, No. 3, 1997: 305).

10 For Foucault, biopolitics was an 'influence on life that endeavours to administer, optimise, and multiply it' (Michel Foucault, *The History of Sexuality*, trans. by Robert Hurley, London: Penguin, 1998: 137). Derek Gregory contends that it was achieved through a set of regulatory systems (insurance, public health, welfare programmes, etc.) that emphasised wellbeing and control of the body and the society as a whole (Derek Gregory, 'Vanishing Points', *Violent Geographies*, Derek Gregory and Allan Pred eds., NY: Routledge, 2006: 206).

11 According to Westendorff, 'Self-build or self-help housing is the product of a range of activities leading to the design, construction, maintenance and management of the physical structure and immediate surroundings of permanent shelter for human beings. Self-help housing also includes renovations, alterations or adaptations of existing buildings including tenements, industrial spaces, or other structures that have not been occupied for lengthy periods and whose new residents or others working with them undertake the improvements' (Westendorff, 2009: 38).

12 Interview with Jae-soon Yoo (21 August 2014).

13 Interview with Nun Magdalena (26 August 2014) who lived and worked as a garbage collector in Nanjido for 3–4 years in the late 1980s.

14 This project was called 'greenerisation'; ironically, during the 1980s under the dictatorship of the Fifth Republic, a national policy to conscribe dissident college students was also called 'greenerisation'. From the interview with Yong-soo Kim

(16 May 2014) who was in charge of Nanjido's waste treatment and the habitants relocation during the late 1990s and early 2001. Here we can see a historical trace of the association between environmental and ideological cleansing.

15 Byung-cheon Choi, 'Nanjido Report', *New Family*, Vol. 369, May 1987: 34–42. Newspaper reports on the lives and living conditions of Nanjido people from 1980 to early 1990s similarly described their lives as human-interest survival stories, seen from an outsider's perspective. See *Dong-A Ilbo* (15 July 1980); *MK Business News* (19 September 1981); *Dong-A Ilbo* (23 July 1984); *Kyunghyang Shinmun* (9 January 1990); and *Kyunghyang Shinmun* (13 December 1990).

16 The shared shower facilities (125 m^2 for men and women respectively) were designed large enough for 40 people to shower simultaneously. See *Dong-A Ilbo* (27 April 1989); *Kyunghyang Shinmun* (27 July 1989); and *Hankyoreh* (28 April 1989).

17 Interview with Yong-soo Kim (16 May 2014).

18 Gilles Deleuze and Félix Guattari, *Anti-Oedipus*, trans. by Robert Hurley, Mark Seem and Helen Lane, London: Continuum, 2004 [1972]: 23, 33–36.

19 Claude Lévi-Strauss, *The Savage Mind*, IL: University of Chicago Press, 1966 [1962].

20 Deleuze and Guattari, 2004: 7–8.

21 Ibid.: 33–36.

22 Westendorff, 2009: 37–38; ESCR, 2003.

23 Ibid., 2009: 38, 55; Miloon Kothari, Statement of the Mr. Miloon Kothari Special Rapporteur on adequate housing as a component of the right to an adequate standard of living to the Commission on Human Rights, Fifty-ninth session, Agenda Item 10 (4 April 2003). In a discussion on the full subjectivity, Achille Mbembe mentions the 'rights to health, education and a functional economy' as basic conditions for full civil, political and socio-economic rights (Achille Mbembe, 'Necropolitics', *Public Culture*, Vol. 15, No. 1, 2003: 13).

24 Interview with Rev. Kyung-Whan Chang (13 August 2014).

25 On one of the biggest floods in the Nanjido Landfill area, see *Hankyoreh* (15 September 1990). As for fire in the Nanjido Landfill and residential area, see *Kyunghyang Shinmun* (28 April 1992); *Dong-A Ilbo* (31 March 1993); and *Hankyoreh* (14 February 1996).

26 Interview with Yong-soo Kim (16 May 2014).

27 The landfill fire was due to the unsanitary landfill's lack of infrastructures for adequate landfilling, e.g. methane gas and leachate treatment systems.

28 *Dong-A Ilbo* (17 January 1990). Regarding the landfill workers' exposure to toxic chemicals, see *Kyunghyang Shinmun* (9 October 1996; 26 September 1991).

29 *Kyunghyang Shinmun* (30 January 1987).

30 Interview with Rev. Kyung-Hwang Chang (13 August 2014) and Nun Magdalena (26 August 2014).

31 *Dong-A Ilbo* (28 June 1993).

32 'The small stream, which we called "poop river", and the bridge formed the borderline. Here was hell and there was heaven. The children of this zone could not play with the children from that zone. To the parents of that region, we were just a group of ragpickers, so they did not permit their children to get along with the ragpickers' children'. Interview with Nun Magdalena (26 August 2014).

33 On after-school activities in the Nanjido Landfill residence, see Ju-hye LeeYou, 'Painting World in Nanjido Landfill', *Saemteo* Vol. 32, No. 1, January 2001: 59. For nurseries run by religious organisations, see Hyun Kang, 'Waiting for Love', *New Family* 386, December 1988: 102; *Hankyoreh* (12 November 1991).

34 Interview with Nun Magdalena (26 August 2014).

35 Westendorff, 2009: 56–61.
36 Interview with Nun Magdalena (26 August 2014).
37 Ibid.
38 *Dong-A Ilbo* (19 February 1994).
39 Seoul City, 'The Processes of the Relocation of the Nanjido Collective Housing Complex Residents' (document) (August 2001) and the interview with Yong-soo Kim (16 May 2014).
40 Twenty-five formal official public meetings were held from 1998 to 2001. Prior to the decision to build the 2002 FIFA World Cup Korea/Japan main stadium in the Sangam-dong area, the City relocated 760 families to public leased apartments from November 1993 to May 1997. It also devised a plan to move an additional 199 families in June 1997 (49 families to the Shintri area and 147–150 families to the new Sangam-dong district). While the World Cup Park construction was in progress, the remaining residents continued negotiating with the City over their temporary residence before moving into the new Sangam-dong apartments. See Seoul City (document) (August 2001). Interviews with Yong-soo Kim (16 May 2014) and Rev. Kyung-Hwan Chang (13 August 2014).
41 Interview with In-hwa Jung (14 May 2014), former Seoul City manager of the waste vehicles in Nanjido Landfill.
42 *Hankyoreh* (26 December 1992).
43 *Hankyoreh* (19 June 1993).
44 Etymologically, 'un' suggests anomaly; '*un*employment' is a name for a markedly temporary and *abnormal* condition. The society therefore casts employment as a key—*the* key—to the resolution of the issues of socially acceptable personal identities, secure social positions, individual and collective survival, social order and systemic reproduction (Bauman, 2012 [2004]: 10).
45 Ibid.: 5.
46 Interview with Rev. Kyung-Whan Chang (13 August 2014).
47 There is no official record of the garbage collectors' income mainly because most people did not pay nor report their income taxes. Newspaper reports on their income also vary: 30,000–200,000 KRW (30–200 USD), *Dong-A Ilbo* (15 July 1980); 500,000–700,000 KRW (500–700 USD), *MK Business News* (19 September 1981); 120,000–300,000 KRW (120–300 USD), *Dong-A Ilbo* (23 July 1984); and 600,000–700,000 KRW (600–700 USD), *Hankyoreh* (26 December 1992). In the mid-1980s, a mid-level public servant's monthly income was approximately 250,000–400,000 KRW (250–400 USD). In the early 1990s, the monthly income of an early career salaryman (e.g. teacher) was approximately 500,000 KRW (500 USD).
48 *Dong-A Ilbo* (23 July 1984).
49 Interview with Rev. Kyung-Hwan Chang (13 August 2014).
50 William Rathje and Cullen Murphy, 1992: 33.
51 *Dong-A Ilbo* (15 July 1980); and *Dong-A Ilbo* (23 July 1984).
52 Interview with Jae-soon Yoo (21 August 2014).
53 Interview with Nun Magdalena (26 August 2014).
54 Freeganism is a neologism derived from 'free' and 'vegan', which also has the meaning of free gaining, an extreme anti-consumerist and anti-capitalist ideology. It began in New York in the mid-1990s and spread throughout major wealthy cities around the globe, including London, Berlin and Paris. Freegans' practices were extended to squatting, sharing and urban gardening, etc. (Jieun Shin, 'Waste and the Reconstruction of Urban Consumer Culture', *Journal of Modern Social Science*, Vol. 17, 2013: 17–38).
55 Walter Benjamin, *The Writer of Modern Life*, Cambridge, MA: Harvard University Press, 2006: 108.
56 Ibid.

57 One year before writing 'Le Vin des chiffonniers', Baudelaire published this prose description of the figure (Ibid.; Baudelaire, *Oeuvres*, Vol. 1: 249 ff. Benjamin's note).

58 Bauman, 2012 [2004]: 27.

59 Zabbaleen (زبالين Zabbalīn) literally means 'garbage people' in Egyptian Arabic and refers to 'garbage collectors' who serve in the Mokattam village in Cairo. The Mokattam village, known as a garbage city with a population of approximately 60,000, is assumed to be the largest city of informal garbage processing (Ragui Assaad, 'Formalizing the Informal? The Transformation of Cairo's Refuse Collection System', *Journal of Planning Education & Research*, Vol. 16, No. 2, 1996: 115–126).

60 Rathje and Murphy, 1992: 40.

61 Andy Merrifield, following Henri Lefebvre, defines 'lived space' as 'the space of everyday experience' (Andy Merrifield, *Henri Lefebvre*, London: Routledge, 2006: 109). Lived space, in Marc Purcell's words, 'represents a person's actual experience of space in everyday life' (Mark Purcell, 'Excavating Lefebvre', *GeoJournal*, Vol. 58, No. 2, 2002: 102).

62 Agamben's theory on the space of exception begins with the camp where life can be stripped of subjectivity and rights, and exposed to death—bare life. See Agamben, 2005.

63 Penny, 2010: 5.

64 Ibid.: 17.

65 In the discussion of the stench of the poor in the nineteenth-century French society, Alain Corbin mentions 'sailors' as one of the major groups along with ragpickers and factory workers. See Corbin, 1994: 147.

66 Interview with Nun Magdalena (26 August 2014).

67 Interview with an anonymous interviewee and Jade Keunhye Lim (12 April 2014), the head of the exhibition team of the National Museum of Modern and Contemporary Art, Seoul (former senior curator of the Seoul Museum of Art).

68 Penny, 2010: 3; Ronit Lentin, 'Introduction-Thinking Palestine', *Thinking Palestine*, Ronit Lentin ed., London: Zed Books, 2008.

69 Gregory, 2006: 4.

70 Penny, 2010; Lentin ed., 2008: 7.

71 Frantz Fanon argues that there is a principle of reciprocal exclusivity in the colonial city or town (Penny, 2010: 17; Frantz Fanon, *The Wretched of the Earth*, London: Penguin Classics, 1990 [1963]: 30).

72 Interview with Dr Woon-soo Kim (22 April 2014).

73 Samhan refers to three castle-town-states consisting of Mahan, Jinhan and Byeonhan, approximately 3–2 centuries BC Korea. Records on *sodo* are found in the *Book of the Later Han* and *Records of the Three Kingdoms*.

74 *Oxford Dictionary*. Current senses, referring to political refuge or to an institution for the mentally ill, date from the eighteenth century. It also has the meaning of 'shelter or protection from danger' but without political or institutional implications.

75 Interview with Jae-soon Yoo (21 August 2014).

76 According to the testimony of Jae-soon Yoo, who published the non-fiction novel *Nanjido People* (Seoul: Geulsure Publishing, 1989), which is based on her own experiences in Nanjido Landfill, the landfill people did not welcome the political protesters who came into Nanjido Landfill looking for a hiding place rather than for work as garbage collectors.

77 Matthew Gandy, 'Zones of Indistinction', *Cultural Geographies*, 13, 2006: 502 and Richard Ek, 'Giorgio Agamben and the Spatialities of the Camp', *Geografiska Annaler: Series B*, Vol. 88, No. 4, 2006: 365.

78 Bauman distinguishes asylum seekers and economic migrants from refugees; he defines the former groups as the new power elite of the globalised world, while the latter as the more clearly visible targets for surplus anguish. See Zygmunt Bauman, *Society Under Siege*, Cambridge: Polity, 2002.

79 Bauman, 2012 [2004]: 70–71, 87. The construction of more prisons, the increase in offences punishable by imprisonment, the 'zero tolerance' policy and harsher and longer sentences are best understood as the many efforts to rebuild the failing and faltering waste disposal industry—on a new foundation more in keeping with the novel conditions of the globalised world.

80 We can extend the fear of uncertainty caused by border crossings to the matter of social change in general (Bauman, 2012 [2004]: 115); Alberto Melucci argues, 'When contemplating change, we are always torn between desire and fear, between anticipation and uncertainty' (Alberto Melucci, *The Playing Self*, Cambridge: Cambridge University Press, 1996: 43). *Uncertainty* is also analogous to what Ulrich Beck calls *risk*, which is awkward and vexing, but the stubborn and undetachable companion of all anticipation.

81 Butler, 2012: 141–142.

4 From landfill to post-landfill park

At the end of 1992, Nanjido Landfill was officially closed, and it was followed by two different phases: the stabilisation[1] of the former landfill and the building of Nanjido Post-Landfill Park (The World Cup Park). The latter was not initially planned to follow the site's stabilisation, and the decision to revitalise the landfill area was made only after Seoul City appointed the Sangam-dong area across from the closed Nanjido Landfill to be the site of the 2002 FIFA World Cup main stadium. The Nanjido Post-Landfill Park occupies 3,454,545 m² of land, including the former Nanjido Landfill, the collective housing complex site and additional land along the Han River. The Park consists of five sub-parks: the Noeul Park (Landfill 1), Haneul Park (Landfill 2), Peace Park (Landfill 3), Nanjicheon Park (the collective housing complex site) and Nanji Han River Park, which are equipped with facilities for leisure activities (Figures 4.1, 4.2, Table 4.1). This chapter first examines the social background of park-building, model cases for Nanjido Post-Landfill Park and the technical processes of the landfill's stabilisation and reclamation before examining how Nanjido Post-Landfill Park embodies the 'global style' of parks.[2] The last part explores the meanings and influences of 'environmentalism'[3] in the context of the global market economy. By interpreting environmentalism in the new economic system as an ideology that uses concepts and practices related to the conservation of the natural environment as an exchange value, I argue that Nanjido Post-Landfill Park produces and functions as exchange value in light of urban redevelopment. I also show that the building of Nanjido Post-Landfill Park continued the principles of urban cleansing and the politics of sanitisation.

The regeneration, stabilisation and reclamation of the landfill were intended to remove the pollutants of the closed landfill and to prevent further damage to the natural environment and/or the health of the population. These steps were also taken to revitalise the landfill to an unpolluted state similar to that of the natural environment that predated it. Technologically, removing leachates and treating gas and odours through biological and chemical methods mainly stabilised the Nanjido Landfill. The government, landscape architects and environmental experts, particularly those involved with engineering, have assessed that Nanjido Landfill's stabilisation owes

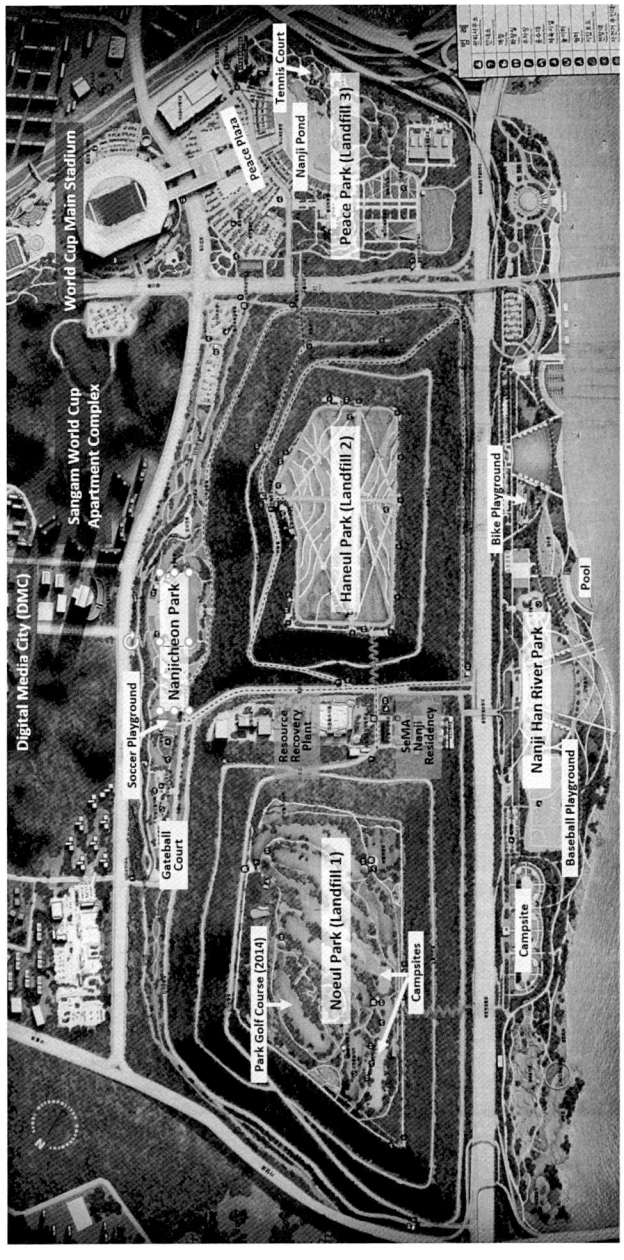

Figure 4.1 Nanjido Post-Landfill Park (The World Cup Park) in Seoul. The map indicates five sub-parks and sites for leisure activities, such as campsites in Noeul Park and Nanji Han River Park. On the north-eastern side is the World Cup main stadium and on the north of the Park is the Sangam-dong New Town. c. 2002. Map © Seoul City

Figure 4.2 Top of Haneul Park (Landfill 2)—grassland and pathways for free activities. Photo by Sook-ja Kim © Seobu Park & Landscape Management Office, Seoul City

Table 4.1 Size and use of Nanjido Post-Landfill Park

Nanjido Post-Landfill Park	Size	Use
Total	3,454,545 m²	N/A
Noeul Park (Landfill 1)	347,107 m²	Leisure activities Golf → camping and park golf
Haneul Park (Landfill 2)	165,289 m²	Eco Park (educational purposes): a space for free activities
Peace Park (Landfill 3)	462,810 m²	Leisure activities e.g. picnic, sports and special events
Nanjicheon Park(The Nanjido Landfill collective housing complex site)	396,694 m²	Leisure activities e.g. picnic and sports (soccer)
Nanji Han River Park	776,860 m²	Leisure activities e.g. picnic, camping and sports (e.g. baseball)
Landfill 1, 2 slope	1,305,785 m²	Eco Park

Table by the Seoul Institute

its success to the fact that the pollutants have been cleansed to meet safety measures without accidents or side effects.[4]

Considering that the landfill's stabilisation is the removal of any source of environmental contamination and potential health threat, it is comparable to the sanitisation of the urban environment, which reached its peak in South Korea in the 1960s and 1970s. While the Nanjido Landfill maintained the cleanliness of Seoul's urban space as a municipal waste management facility during the industrial and post-industrial periods, the stabilisation of the landfill, which had eventually become a mass of waste itself, was the post-landfill era's new form of sanitisation. In this new phase, landscape architects as well as politicians implemented urban sanitisation by transforming the previous era's abandoned or deteriorated structures into new ones that may represent the values of the new era: environmental awareness and cultural abundance. The endeavour aimed to create a sanitary city differently from efforts of the previous period; while the government of the past had divided the city space into waste (the inappropriate) and non-waste (the appropriate) through zoning or bordering, the governing body of the new era dexterously scattered or hid the waste to make it invisible within the urban space. Urban regeneration, likewise, accompanied the advanced technique of hiding the waste. In short, the Nanjido Landfill's stabilisation was intended to physically cleanse pollutants that had accumulated in the land, water and air throughout the landfill period. Stabilisation would ultimately regenerate the natural environment so that native floras and faunas would revive and the environmental safety levels would be fit for the public's[5] leisure activities.

Robert Weller, in his book titled *Discovering Nature: Globalization and Environmental Culture in China and Taiwan* (2006), argues that globalisation and environmental awareness simultaneously emerged as dominant discourses at the turn of the new millennium; globalisation entails planetary pollution, and it subsequently and ironically increased environmental awareness.[6] Here, I examine the meaning of environmental awareness and environmentalism during an era of new globalised economic systems, artificial nature, or the landscape park, and its relationship to humanity. In discussions of environmentalism, the following concepts frequently emerge in support of the symbolic practice of park-building: regeneration, nature, leisure, wellbeing and the public. Examining the meanings behind park-building, the regeneration of the natural environment and the leisure activities in the park illuminates connotations of 'the public' that both the city and environmental organisations revealed. Analysing the leisure activities in Nanjido Post-Landfill Park, in particular, will address the idea that 'the public' includes or excludes certain socio-economic groups.

As Weller argues, 'the globalisation of nature', represented by nature tourism, is the epitome of the anthropocentric attitude toward nature and one of the class-based phenomena that only the middle class, people with financial leeway, leisure time and access to transportation, can enjoy.[7] The shift toward such leisure activities has become more significant in the new economic society, which sees a growing gap between the rich and the poor; for example, the debate on the construction of a golf course in Nanjido Post-Landfill Park not only disputes efficient land use during the landfill's stabilisation and reclamation processes, but also questions the boundary of potential users—the public. This chapter will show how the discourse on leisure activities for the public's wellbeing often excludes the lower income classes.

This analysis of Nanjido Post-Landfill Park under neoliberalism and its urban environment lays out the ways in which government or corporate or individual entities endeavour to enhance their wealth and power by using the ideology of environmentalism in a manner similar to how previous regimes had utilised exploitation of the ideology of sanitation for political power and economic development.

The building of Nanjido Post-Landfill Park

Post-landfill plans

The idea of park-building on the Nanjido Landfill site dates back to 1985 when the underground landfill was almost complete. Seoul City had been preparing for overground dumping and considering the post-landfill period. At that time, the City had planned to continue dumping garbage until the landfill had filled up the wetland and reached ground level at 50–70 m in height before announcing its closure.[8] The next step was to transform the closed landfill site into a public park equipped with sporting facilities, including a golf course and a baseball field and so forth. However, research on the park-building's environmental impact (by the then Environmental Administration) had annulled the entire plan; the research concluded that the overground landfill would have a high risk of collapse and leak leachate and methane gas that could result in a natural fire or explosion, particularly because of the mixed waste dumped in its original material form.[9] As this research opposed the overground landfill itself, it automatically annulled the post-landfill park-building plan, too.

Still, without an alternative plan for waste management, Seoul City officially continued to add waste to the overground landfill for the next 7 years, while the City and private organisations also proposed various post-landfill development plans. After the 1988 Summer Olympics in Seoul in particular, the post-landfill park development plan re-emerged out of increased recognition of the city's need for cleanliness. Advanced technologies to control pollutants also supported the feasibility of the immediate development, which enabled the City to have confidence in the practical possibility of the closed landfill's immediate development. All the same, the policy makers were also aware that it would be better to wait over 10 years for the land to settle before developing the site.[10]

At the approach of the landfill's closure in the early 1990s, post-landfill plans re-surfaced specifically in connection with housing development, and after the mid-1990s, this issue practically became intertwined with the Sangam-dong regional development. In 1991, Seoul City suggested concrete plans: 1) housing development on the landfill site as the first option; 2) park-building as the second; and 3) natural decomposition (leaving the landfill site untouched for about 40–50 years before pursuing housing development) as the third.[11] Meanwhile, large corporations, such as Samsung, proposed to develop the former landfill site into a large-scale teleport town.[12]

After examining the waste dump relocation's environmental impact and the housing development's economic effect, the City decided to proceed with the landfill's stabilisation and reclamation first. The resolution left the park-building as a high possibility for the future, but the final decision had not been made. It was not until the City designated the Sangam-dong area as the site for the 2002 FIFA World Cup main stadium (1998) that it confirmed the Nanjido Post-Landfill Park (1999).[13] The official announcement accelerated the Sangam-dong New Town Development Plan (1999), which was eventually expanded to the Sangam-dong New Millennium Town Plan (2000). This included the development of the Digital Media City (beginning in 2002) in the area.

Likewise, the post-Nanjido Landfill plans had oscillated between housing development and park-building for more than a decade. The post-landfill preparation's link to the building of the 2002 World Cup main stadium and the Sangam-dong redevelopment plan justified the landfill's transformation into a park based on a belief that it would bring about potential brand value and environmental value to the region.

Models of the post-landfill park

The transformation of the industrial age's closed landfill into a park began as early as the 1980s and intensified in the 1990s. Seoul City used model cases of other cities as references for the construction of Nanjido Post-Landfill Park.[14] Amongst the cases of the West, Seoul City selectively referred to different aspects of several post-landfill parks, including Stockley Park in London, Byxbee Park in California and Sydney Olympic Park. Freshkills Park, though not used as a reference, is similar to Nanjido Post-Landfill Park, since it helps our understanding of landfill regeneration characteristics under twenty-first-century socio-economic circumstances.

Stockley Park, an approximately 370-acre tract of land in the London Borough of Hillingdon, had been used for gravel workings and was later backfilled with municipal refuse. This case shows an early incidence of a former landfill's transformation into a business park combined with a landscape park.[15] Although the project was first proposed in 1981, the developer Stockley Park Consortium Ltd. initiated it in 1985. Arup Associates was selected as the chief developer, and seven other architectural firms, including

Norman Foster & Partners and Skidmore Owings & Merrill, participated in the design process.[16] The Park successfully let and sold the estate to corporations such as Apple, while the design gradually changed its focus to environmentally conscious landscaping; for example, Arup Associates proposed measures to promote biodiversity, including a gateway sculpture that would also function as a bat roost.[17] The strategy of re-using on-site materials is one of the most notable characteristics of Stockley Park's construction; transferring refuse from the Business Park area to make a new topography on what was previously flatland created the landform for the Country Park, including the 18-hole golf course.[18]

Byxbee Park in Palo Alto, California is one of several San Francisco Bay area parks that began as landfills. The landfilling started in the 1960s and the garbage was dumped in a non-systematic way in several different locations. When the City of Palo Alto decided to convert the landfill site into a park, the landscape architects Hargreaves Associates and the artists Peter Richards and Michael Oppenheimer led the project (1988–1992). Out of the total 126 acres of land, a 29-acre park opened in 1991. This project's technologically driven environmental achievement was the underlying system of methane collection that helps provide power to the city. Later, Seoul City applied that system to Nanjido Post-Landfill Park.[19] Above all, Byxbee Park shows a unique collaboration between art and landscape design that reclaims the ecosystem; the piece titled 'Pole Field', which consists of a series of 72 uniform perch poles, provided sites for birds to rest.[20]

The building of the Sydney Olympic Park, part of the redevelopment of Homebush Bay (a suburb of Sydney) at the time of the Sydney 2000 Summer Olympics, was the most similar case to Nanjido Post-Landfill Park in the sense that a municipal event spurred the transformation of the landfill into a park. Historically, extensive landfilling had begun in the Homebush Bay area since the 1890s, and industrial waste dumping had continued into the early 1980s. The reclamation of land, which had proceeded mostly during the 1960s and 1970s, had progressed at a slow pace. Once Sydney won the right to stage the 2000 Olympics in 1993, however, regional redevelopment, including the landfill site reclamation and park-building, accelerated.[21] Despite the debate on the environmental issues surrounding the hasty clean up during the Olympics—wetting down and moving the waste in bulk without sealing and bagging it—the region is still known for its transformation into a parkland with sports facilities, musical and cultural events and a residential area.[22]

Freshkills Park (opened around 2010), the 8,900,000 m^2 of land (approximately 2.5 times larger than Nanjido Post-Landfill Park) located in Staten Island, New York, is the world's largest example of a post-landfill landscape park. The Fresh Kills Landfill opened as one of public official Robert Moses' urban planning measures in 1948 during the United States' most productive industrial age, and it operated until 1997. In 2001, Field Operations led by James Corner was selected as the first-prize winner of the Fresh Kills

Landfill to Landscape Design Competition, City of New York. A decade after the landfill's closure in 2006, park construction began. The first part of the project opened in 2010, but the Park's development will continue until its expected completion in 2035. The Freshkills Park is designed to accommodate educational, sporting and cultural facilities, and circulation paths for bicycles, pedestrians and equestrians will connect its various spaces[23] (Figure 4.3).

Although Nanjido Post-Landfill Park was unable to refer to the Freshkills Park as a model because Nanjido's groundwork was almost already completed by the time Freshkills Park's plan had been initiated, Freshkills Park's characteristics are significant to the notion of the post-landfill park of the twenty-first century, including Nanjido Post-Landfill Park. Applying the most advanced technologies of landfill stabilisation and reclamation that least affect the environment, the Freshkills Park aims to revitalise its native ecosystem and encourage the public's leisure activities by providing educational, sporting and cultural facilities.

As Nanjido Post-Landfill Park was planned in the late 1990s, Seoul City used these earlier cases in which closed landfills had been transformed into parks, except for Freshkills Park, as a reference. The City adopted specific practices from each case to raise the brand and environmental value of Nanjido Post-Landfill Park as well as to build a new era under new socio-economic circumstances; for example, Seoul City took the successful

Figure 4.3 Freshkills Park, New York. © Alex Maclean and NYC Parks department

example of a golf course from Stockley Park (the City might have also used its preceding Business Park plan as a model for Sangam-dong's teleport town, which combines the Digital Media City with the Park). The City took the idea of combining the Post-Landfill Park with art from Byxbee; in Nanjido's case, however, the fusion with art did not enable the works of art to engage with the remediation of the ecosystem. Instead, the works of art functioned to increase the cultural value of the park. In light of the Sydney Olympics, Seoul City confirmed the advantage of comprehensive regional redevelopment, which would integrate residential, commercial and leisure facilities, while raising the total brand and real-estate value. In terms of the Park's function itself, the Freshkills Park continues to be a compatible case due to its concern for the environment and public's wellbeing.

Nanjido Landfill's regeneration[24]

The completion of Nanjido Landfill's overall stabilisation and reclamation was expected within a relatively short period of time because food waste constituted 46% of the waste amassed during its landfill period (1978–1992), and waste of rapid decomposition was dumped after the landfill's closure in 1992. Although waste with slow decomposition rates, such as plastic, rubber and leather, formed 24% of the total waste, the City expected the completion of 85% of the land stabilisation and reclamation by 2050.[25]

In a broader sense, the landfill's stabilisation and reclamation processes involved treating the leachates, gas and odour and planting throughout the closed site. Nanjido Landfill's stabilisation mainly consisted of leachate and gas treatments with a focus on managing methane and other gases. While leachate treatment is meant to purify water and gas control is intended to purify the air, controlling underground leachate and methane gas is ultimately designed to regenerate the land quality. Practically, the land subsidence occurs in tandem with the removal of leachate and the decrease in the generation of gas.

Even though deodorisation is part of the stabilisation processes, researchers did not include the degree of odour as a measurement of the landfill's stabilisation. First, this shows that researchers thought the odour could be cleared away after certain gases were properly removed. Second, it indicates that they did not regard odour, being sensible yet intangible, to be as fatally harmful to the land or the health of human and non-human beings as certain other gases. Third, it suggests the difficulty of specifying that the exact cause of the odour was the unsanitary landfill; the odour stemmed not only from the landfill, but also from neighbouring sewerage facilities and streams. Although the stench of an odour is stronger than that of other elements, stench itself was not necessarily the cause of harm but a symptom of other forms of pollution.

Amongst the four processes, leachate and gas treatments were the main stabilisation practices because they would eliminate harmful pollutants. On the other hand, odour management and planting were more related to the reclamation, or building of Nanjido Post-Landfill Park, because they masqueraded or neutralised the detested sensory (olfactory and visual) elements that remained.

Detoxification: leachate and gas treatments

As the landfill's stabilisation was intended to remove the pollutants from the site, practical stabilisation methods included the construction of leachate and gas treatment facilities. For leachate treatment, leachate collectors and pipelines had been installed to remove the leachate from the landfill and impervious water-blocking walls had been established to prevent further contamination of the Han River. For gas treatment, numerous gas collectors and pipelines were set up to remove the collected gas from the landfill.

First, according to the investigations conducted before the technical leachate treatment, the pollution rate of the leachate was particularly high in Noeul Park (Landfill 1) where industrial sludge remained. Along with ammoniacal nitrogen (NH_3-N), the volume of heavy metals (e.g. chromium, copper and lead) was substantial in that area. Technically, since non-biodegradable materials formed a large portion of the wastewater, they employed physical and chemical treatments rather than biological ones as the primary methods for treatment.[26]

Prior to establishing the leachate and sewage sludge treatment facilities, it was essential for the leachate treatment construction to cover the surface of the two landfill mounds with layers of topsoil, HDPE (High-density Polyethylene) and other fabrics to prevent further leachate leakage and rainwater absorption. The landfill slope produces approximately 75% of all the leachate in the Nanjido Landfill, while its base produces 25% and its exterior surface 0.3%. As those involved could not install the blocking material of HDPE on the slope because of the plants in cultivation, however, they built alternative structures such as sewage paths to allow the water to flow down the slope without being absorbed into the landfill.[27]

Leachate treatment is related to both the purification of polluted water and the landfill subsidence that gradually occurs as the leachate leaves the landfill. Since the landfill contains a multitude of different substances in different areas—for example, Noeul Park (Landfill 1) has the highest concentration of industrial sludge, whereas the large slope area toward the south of the Han River side is full of municipal solid waste[28]—the City had to fill up certain sections with extra soil to reinforce areas that were more likely to sink rapidly, thereby solidifying the overall land mass for a balanced subsidence.[29]

Second, an equally important stabilisation process was the gas treatment. As organic materials break down over time, anaerobic microorganisms help the process, and this generates the landfill gas, which consists of methane, carbon dioxide, nitrogen, oxygen, water and dozens of VOCs (Volatile Organic Compound). Methane, in particular, causes a high risk of fire and explosion, hydrogen sulphide and ammonia cause odour, and certain VOCs are hazardous to health and deter the growth of plants. Collecting the landfill gas through gas collecting systems installed all over the landfill site played a major role in stabilising the region (Figure 4.4). The methane gas collected through these systems went to the gas purification system (near the Resource Recovery Plant) to be treated and processed into a source of fuel that would heat residential and commercial areas in the North Sangam-dong region.

The most important criterion for the landfill's successful stabilisation and reclamation, or transformation into a 'land of opportunity', is the extent that VOCs have been reduced. Therefore, the Seoul Institute investigated VOCs in and around the closed Nanjido Landfill site four times (four seasons) in 2000. It measured the impact of the VOCs on the health of adult human beings exposed to the atmosphere under the conditions of three different virtual scenarios (Worst, Moderate and Lowest Inhalation Exposure Scenarios). The research results show that, except for the worst-case scenario during the first half of the year, the other groups turned out to be safe during all four seasons according to the risk assessment standard[30] of VOCs.

As the leachate and gas treatment facilities installed in the Nanjido Landfill show, the landfill stabilisation first removes toxic elements by purifying the materials then blocks the pollutants from leaking outside the landfill site. In other words, it is both a process that rehabilitates the deteriorated (inappropriate) land to the clean (appropriate) urban territory, and a process that confines the toxic elements to a certain boundary and separates them from other regions to avoid contaminating the surrounding environments. While the overground landfill area's transformation into a park merges into the city space, the underground leachate blocking wall, another physical and symbolic border, will remain to prevent toxic liquids from leaking out even after 2029, the expected year of completion for the first phase of the landfill site's stabilisation (excluding that of waste with a slow decomposition rate).

Aestheticisation: deodorisation and planting

In the Nanjido Landfill's regeneration processes, leachate and gas treatments are regarded differently from odour management and planting. While the former two eliminate the causes of pollution, the latter two cover or conceal pollutants and the polluted landscape. Seoul City was particularly concerned with deodorisation and planting as they were more tangible (olfactory and visual) than the other two. The City had to address these elements to meet the standards of environmental appropriateness for the 2002 FIFA World Cup.

First, deodorisation, amongst other regeneration processes, carried a lot of weight for the 2002 FIFA World Cup since the stench could leave a bad

Figure 4.4 Methane gas pipelines installed all over Nanjido Post-Landfill Park. 2014. © Jeong Hye Kim

impression on the FIFA committee judging the area's appropriateness for the international sporting event. With that in mind, Seoul City strongly pushed for the completion of deodorisation in and around the landfill site by the FIFA committee's on-site inspection in November 2001.[31] The odours of the Nanjido Landfill, especially those hazardous to health, were mainly related to the landfill gases. The strongest measured odour was found on the top surface of the landfill before gas collection began, whereas the plants on the landfill slope had masked the stench, making the odour relatively low. Other causes of nauseous odours included those from: 1) the Food Composting Facility (5 km from the Sangam-dong area); 2) the agriculture and fishery market east of the landfill; 3) the livestock excretion used as compost in the then farmland[32] located on the southern part of the landfill; 4) the sewage of nearby streams; and 5) the Nanji sewage treatment facility (waste water pumping facility) at the corner of the landfill site.[33] Although the main cause of the odour in the landfill area was the mixture of decomposing garbage, the landfill's regeneration engineers managed stenches from other sources as well to improve the conditions in areas neighbouring the World Cup main stadium. Despite the Food Composting Facility's distant location from the Nanjido Landfill and the World Cup main stadium, it generated considerably strong effluvia, so engineers re-operated the existing soil deodorising devices and installed an air curtain at the facility's entrance. City experts suggested that the Facility stop operations during the World Cup games in case it continued to generate stench.[34] To remove the fetid odour from the riverside farmland, the City urged the farmers to replace the compost with chemical fertiliser and compensated them for the extra cost of the chemicals. Still, the City let them finish ripening the compost by the summer of 2001, before the FIFA inspection in November of that year.[35] As such, the odour's diverse causes and extensive permeation of physical boundaries led the City to implement deodorisation widely enough to control the entire Sangam-dong area.[36]

Second, planting was somewhat a stabilisation process and somewhat an aestheticisation factor for Nanjido Post-Landfill Park and surrounding regions. On the one hand, planting trees bolstered the land and helped land stabilisation. On the other hand, it fulfilled the goals of the City's plant ecology recovery project; Seoul City expanded the initial planting plan to incorporate the recovery of the Nanjido region's native natural state, which was also a part of the city-wide project to create a green zone that vertically stretched along the city of Seoul. A project to rehabilitate the botanical environment establishes the infrastructure to re-build the native ecosystem, in the sense of a pre-landfill state of nature, in harmony with the region's overall ecosystem. The biologists primarily used floral regeneration to replace pollutant-resilient species (which constituted 27% of all the plants on the landfill slope before the landfill regeneration) with native plants. They, then, built environments that native faunas could inhabit. For amphibians, they built marshlands with clean flowing water and a ramp way connecting the road to the woods; for birds, they constructed diverse microhabitats with roosting covers, hiding places and plants for food.[37] So far, these habitats have been well managed as initially planned, crediting the post-landfill site as an 'eco-park'.[38] The revival of native floras and faunas progressed in consideration of the City's appointment as host of the World Cup and the football stadium's location; for example, they planted more aesthetic species on the north-eastern slope of Haneul Park (Landfill 2), the main landscape viewed from the World Cup stadium.[39]

Technically, Nanjido Landfill's regeneration was the landfill's stabilisation—the restoration of the unsanitary abandoned land (the inappropriate) to the sanitary urban territory (the appropriate) by means of detoxification technologies. Nevertheless, it was also its reclamation, or an effort to aestheticise the region, which was closely connected to the 2002 FIFA World Cup and the environment around the main stadium. In a broader sense, both movements ultimately intend to block, cover and conceal unsanitary pollutants before they become appropriate to the norms of the urban space. Therefore, the technical aspects of the landfill regeneration can be viewed both as an extension of the sanitary control of the previous eras (spatial separation of waste by bordering) and in line with the environmentalist approach of the new era (making any pollutants or potential threat of disease invisible).

The Nanjido Landfill's regeneration has received positive responses from experts and the public alike, especially because Seoul City left a considerable portion of the city-owned land for public use instead of immediately developing it for commercial purposes. However, it is hard to conclude that the post-landfill site's regeneration was entirely for public purposes as it consequently generated economic gains (creating brand and real-estate values) from the World Cup and the Sangam-dong New Millennium Town development.

The global style of parks

US style and global style of parks

As for its classification, Nanjido Post-Landfill Park is considered a landscape park where people can appreciate nature through its preservation. In light of the global economic circumstances, however, we cannot liken the intent of contemporary landscape parks to the eighteenth-century European yearning for untamed nature and wilderness or to the nineteenth-century American appreciation of nature as a divine presence.[40] Instead, we can first interpret the building of landscape parks since the late twentieth century as a reaction to the environmental degradation caused by over a century of industrialisation—in this sense, the idea of contemporary park-building is analogous to the dynamics that mobilised the US national park movement in the early twentieth century.[41] Second, we can regard the landscape parks that have been built since the late twentieth century as sites symbolically functioning to create an image of the public's increased wellbeing mainly through leisure activities in green spaces and nature tourism. These two traits are components of the global style of parks, which influences present-day landscape parks.

The idea of the love of nature seems to form the foundation for the aforementioned landscape parks. Still, there are subtle but crucial differences to the way people view, manage and utilise nature in each landscape park. This is particularly evident in the differences between the nineteenth-century US style and today's global style.

The US park style is an artificial structure that is meant to represent the 'appreciation of nature', which is based on the transcendental idea about the relationship between nature (as a presence of the divine[42]) and human beings. US park-building also hinges on the 'preservation of nature', especially the protection of natural species in danger of extinction; for example, the bird-watching movement that rapidly swept through the United States since the establishment of the Audubon Society in Massachusetts in 1897 testifies to this. Bird-watching culture had grown in part out of the tradition of local natural history that arose in response to the increasing dominance of the Enlightenment's reconfiguration of the nature-culture distinction.[43] Although the modern bird watchers had somewhat scientifically approached nature with their intentions to conserve (more involved in management than in preservation) and document the diversity of natural species, the primary goal of their activity is rooted in the appreciation and preservation of wilderness, too. Overall, the early US style of parks can be said to 'appreciate and preserve nature' while keeping distance between nature and human beings. Accordingly, the rules and regulations in these early parks discouraged tourists from direct interactions with nature. Any leisure activities, including camping and barbecuing, were illegal in the National Park. That is, nature, in the early US style, existed for our observation, but not for our immersion. Although human beings' attitudes and motivations toward the nature of contemporary park-building are different from those responsible for the early US style of parks, we cannot completely deny that the purpose of the latter park space, the appreciation and preservation of nature, has provided the foundation of the concept behind the landscape parks of today—a distanced appreciation of nature transformed into the consumption of nature's exchange value in a different economic context.

The emergence of the global style of parks is inseparable from the global environmentalism[44] that mainly began to appear in US society after WWII. First, especially in the late 1960s' post-war era, civil rights activists and anti-Vietnam war activists amongst other participants in comprehensive anti-social movements also voiced environmental concerns. They claimed that human-centred industrial development and socio-political conflicts for development resulted in environmental degradation.[45] Second, environmental awareness had also surfaced in the field of biological discipline, represented by Rachel Carson. Carson gave rise to the scientist-activists' argument that industrialisation was destroying the biological resources upon which it depended.[46] The movements of civic and environmental activists and biological scientists constituted the early ideas and practices of environmentalism and called for an end to human-centred development, or anthropocentric developmentalism.

These notions embodied the main structure of contemporary environmentalism, which the United Nations and the International Union for Conservation of Nature (IUCN) officially founded and declared. Global environmentalism recognises various responsibilities from the protection of nature and sustainable resource use to nature tourism. The organisation's guidelines for protected areas include: 1) strict protection (i.e. wilderness areas); 2) ecosystem conservation and protection (i.e. national parks); 3) conservation of natural features (i.e. natural monuments); 4) conservation through active management (i.e. habitat/species management areas); 5) landscape/seascape conservation and recreation (i.e. protected landscapes/seascapes); and 6) sustainable use of natural resources (i.e. managed resource protected areas). Of these six categories, the first three are involved with the protection of nature (i.e. UNESCO's biosphere reserve programme protects genetic diversity in a full range), while the latter three make way for continuous human use of nature.[47] The latter cases hold traces of the anthropocentric idea that the natural environment exists for the sustainable development of human civilisation. Moreover, the terms 'conservation' and 'protection' of nature entail that the intended focus of contemporary environmentalism is the sustainable use and management of natural resources, rather than preserving untrammelled nature, or leaving natural sources untouched.

With respect to the new approaches to the natural environment and the style of parks, combining nature with leisure (e.g. camping, lodging and food services) was the most significant change. Some transportation systems (e.g. freeways, car-ownership and easier access to planes) served as modest means to reach National Parks and Forests, contributing to the extent that people could visit these sites for leisure purposes, too.[48] This shift in the meaning and function of the landscape park has two implications. It represents both humankind's horizontal relationship with nature and its use of nature. In other words, on the one hand, this new approach to the landscape park is based on a relational view of humans and nature, and, on the other hand, it is a new use of the exchange value of nature—the latter being the more frequent case. Likewise, the global style of parks, which is closely related to environmentalism, holds the risk of co-opting the environmentally conscious approach to nature or to the landscape park into the capitalist economic system.

Nanjido Post-Landfill Park as a global style of parks

Nanjido Post-Landfill Park has characteristics of the global style of parks; the Park pursues sustainable use of nature based on nature conservation through active management. Also, the Park was designed based on contemporary environmentalism not only regarding the remediation and conservation of degraded nature, but also regarding its creation of exchange values—'the environmental' and 'the cultural'. The two major goals and functions of Nanjido Post-Landfill Park are environmental awareness and leisure activities (e.g. sporting and artistic practices). Seen through the lens of the exchange value, environmental awareness is related to its own value, or 'the environmental' value, and the leisure activities to 'the cultural' value.

The regenerated Nanjido Post-Landfill Park demonstrates and produces environmental value in its own right and transforms it into economic values in many ways. Group tourism to the Park is one of the undertakings that manifest the Park's environmental value and simultaneously transform it into exchange value. That is, visitors not only experience the regenerated natural environment through tourism but also consume the brand value of the remediated environment as well as the idea of an unpolluted environment in the urban space. Nanjido Post-Landfill Park's environmental value as exchange value is most significant when viewed in the context of the regional redevelopment; while the building of the 2002 FIFA World Cup main stadium sparked and accelerated the Sangam-dong New Millennium Town development and the transformation of Nanjido Landfill into Nanjido Post-Landfill Park, the Park contributed to increasing the brand and real-estate values of the event and the new town. Here, the brand value considerably influences the real-estate value; especially under the new economic systems in place, the exchange values of the environmental and the cultural (leisure activities, e.g. sports and art) are decisive factors of the site's brand value (Figure 4.5).

Figure 4.5 Sangam-dong New Town apartment complex and commercial buildings in the Digital Media City viewed from Nanjido Post-Landfill Park. 2014. © Jeong Hye Kim

Seoul City actively demonstrates the Park's environmental value through various educational programmes. First, the World Cup Park Promotion Centre, located in the east of Nanjido Post-Landfill Park, showcases the historical transformations of the Nanjido region throughout the late twentieth and early twenty-first centuries. The chronology from the pre-landfill to landfill to post-landfill park periods demonstrates how the City restored the polluted landfill to the natural state of its pre-landfill era, by transforming the garbage dump into the environmentally appropriate space of a park. The audience absorbs the dichotomous conception of the clean (the appropriate) and the polluted (the inappropriate) by consuming these indirect experiences. Second, the Mapo Resource Recovery Plant located at the foot of Noeul Park (Landfill 1) represents the rehabilitated post-landfill park's vision for the sustainable use of materials, which would create a sustainable human environment. The Plant provides educational programmes on its automatised recycling systems to a wide spectrum of people from kindergarten kids to foreign experts who may adopt the advanced technology of the waste treatment systems.[49] The programmes practise promoting and marketing the waste management technology and systems along with the idea of regeneration. Likewise, the City-led educational programmes focusing on the environmental regeneration of the post-landfill park, which enables sustainable use of natural resources, potentially reinforces anthropocentric views and an assimilation of the importance of its exchange value rather than its environmental value.

Meanwhile, Nanjido Post-Landfill Park also produces cultural value through diverse leisure activities, including sporting and artistic practices. As a global style of parks, the Park was naturally designed for sporting activities, such as picnics (mostly for kids), camping, park golf[50] and team sports (e.g. soccer and baseball),[51] as well as group tourism.[52] Amongst the diverse activities offered by the Park, the most popular leisure activity in Nanjido Post-Landfill Park is the overnight camping with barbecuing in Noeul Park (Landfill 1) and Nanji Han River Park[53] (Figure 4.6). Campsites are especially busy with groups of families, friends and colleagues in the summertime. In Seoul City, which is rather short of parks that allow lodging and food services, citizens highly welcome Nanjido Post-Landfill Park's inner-city campsites.

As for the artistic aspects, the establishment of the SeMA Nanji Residency (run by the Seoul Museum of Art) as an artistic facility added cultural value to the Park. Since the residency programme focuses on the artistic community rather than on its engagement with the public audience, however, it holds the more symbolic meaning of cultural value than that of the visitors' cultural experience.

Nanjido Post-Landfill Park, which mainly followed the global style of parks, pursues the conservation of the natural environment and simultaneously encourages leisure activities. In other words, Nanjido Post-Landfill Park's status as a landscape park intends to promote the sustainable use of

Figure 4.6 Campsite in Noeul. Photo by Woo-hyun Chun © Seobu Park & Landscape Management Office, Seoul City

nature for the public's leisure activities. The Park also has characteristics significant to the global style of parks since it produces environmental value and cultural value that instantly turn into exchange value, which increases the brand value of both the Park itself and the neighbouring regions. Without a relational consideration of the social ecology, concern for the environmental ecology tends to regard nature as an object for human beings to maintain or exploit, especially based on the logic of the global economy.

Global economy and environmentalism

Anthropologist Robert Weller argues that, in environmental discourses, we must consider the creation of landscape parks as separate from issues of pollution and waste disposal.[54] In my explanation of Nanjido Landfill's regeneration, I also distinguished landfill stabilisation processes (detoxification) from reclamation or park-building (aestheticisation). Regarding Nanjido Post-Landfill Park's situation—where garbage remained on the park site—however, we should also think about park-building in relation to new interpretations of the concept of the 'environment', which is often called 'green' as well.

Building a landscape park is a representation of the new era's 'green ideology'. I call this an ideology because it is based on a belief system that distinguishes the appropriate from the inappropriate using the criteria of cleanliness and the degree of the natural environment's conservation and sustainability. The ideology is what justifies possible problems related to the building of a green environment. Hence, on the subject of park-building, the Seoul City government defined any matter related to Nanjido Landfill, including both discarded materials and human beings, as unsanitary, non-conservational and unsustainable. Since these entities were inappropriate to the new era, the City could justify its decree to annihilate them.

The word 'green' has welcomed an unconditionally positive reception. This has especially been true since global awareness of the environmental crisis emerged in the late twentieth century. Consequently, it tends to create illusory imagery about green-related, environmentally conservational and sustainable practices, while concealing any negative side effects the practices may generate. This issue imbrues Nanjido Post-Landfill Park with ambivalent characteristics, which prevent us from observing the Park from a fixed perspective; for example, whether the exchange value created by the greenerisation of the post-landfill park is perceived as positive or negative depends on the perspective of each social class, or more precisely each income level. Therefore, it is essential to understand the socio-economic circumstances of globalisation and its relationship to environmentalism to untangle the ambivalent reality of Nanjido Post-Landfill Park.

Environmentalism in the global economic system of South Korea

Since the late twentieth century, international academia has carried out studies on the relationship between globalisation and environmentalism,[55] and the discussion intensified during the 1990s when market globalisation had become predominant worldwide. Early debates on environmentalism date back to the 1970s when activists concerned with diverse social matters raised the issue of the environment and political and social scientists examined the quality of a life of material affluence and the role of environmental consciousness within those 'qualitative' factors.[56] Since then, many scholars in related fields began to understand environmentalism as a global concern and developed the study of the relationship between market globalisation and environmentalism in diverse ways. As a result, the topic branched out to encompass varied subjects, including the illusions of environmentalism,[57] the relation between environmentalism and postcolonialism,[58] inclusive environmentalism,[59] ecotourism[60] and specific concerns for the environmentalism in Northeast Asia.[61] These topics mainly emerged in the disciplines of political and social science and scholars aimed to reveal the negative effects of the escalating global market economy and its association with environmental concerns (e.g. commercialised ecotourism). They also tried to find a solution to environmental degradation, particularly in response to problems in developing countries. Overall, they believed that market globalisation caused the human habitat to deteriorate and that environmentalism would solve that problem.

Weller developed a similar argument on the relation between the globalised market economy and environmental awareness. However, unlike the academic approaches above, Weller's approach does not look at the two phenomena as a cause and effect/result/solution relationship; instead, he argues that the two transpired simultaneously. He asserts that developmentalism, taking the shape of land reclamation programmes, continued while practices based on environmental awareness, such as wetland protection plans, rapidly emerged in the late twentieth century.[62] In other words, globalisation has brought about the contradictory practices of market globalisation (represented by the consistent development of the natural environment) and environmental awareness (represented by the protection of the natural environment), and simultaneously disseminated these practices worldwide. In this socio-economic situation, environmentally conscious movements can hardly function as solutions to the large-scale urban development that relentlessly damages the natural environment. On the contrary, the global market system easily co-opts the conserved or revived natural environment and turns it into a source of profit, or exchange value. Based on this idea and Weller's argument, I assert that using concepts and activities related to the conservation of nature as exchange value in the context of the globalised neoliberal economy encapsulates the 'environmentalism' of today.

South Korea entered the new global economic system during the 1990s when concern for environmental degradation began to increase. In 1997–1998, a currency-cum-banking crisis hit South Korea, leading the nation to seek official assistance from the IMF.[63] This economic crisis and the rescuing policies under the new President Kim Dae-jung's administration (1998–2003) signalled that South Korea would decidedly shift toward the globalised neoliberal economic system.[64] Meanwhile, South Korea's representative environmental NGO, the Korean Federation for Environmental Movement,[65] was established in 1993 to protect human lives and the earth from environmental degradation; this objective accelerated environmental movements. In 2002, the Korea Green Foundation was established to create a sustainable society and protect human lives and the environment. These movements unfolded just as environmental consciousness on a global level emerged at the turn of the twenty-first century.[66] Since their establishment, these organisations have functioned to reflect the moral experiences of industrialisation and the consistently increasing production and consumption of today—however, their anthropocentric perspectives on a healthy environment for human lives cast indelible doubt upon their prerogatives. Led by these organisations, South Korea's environmental movements have focused on rehabilitating damaged nature and protecting the natural environment—their primary concern has literally been the environmental ecology, in the sense of the love of nature. Therefore, as long as the polluted environment is revived, the environmentalists will not fret over the co-opted use of a site, particularly when the site is said to serve 'the public'.[67] That is, the environmentalists' concerns are limited to the protection of the natural environment for the wellbeing of humankind (a certain group, vaguely called 'the public'), irrespective of the globalised socio-economic ecology and the relationship between environmental and social ecologies.

Here, it is evident that the co-option of the natural environment with environmental movements is related to the misunderstanding or ambiguous use of the term 'the public'. The conflicts between the government and environmental activists as well as the overall limits of the environmental movements are consequences of the vague definition of 'the public', which is closely related to the society's 'middle class'.

The South Korean middle class, 'the public' and the environmental concern

Although the roots of environmental value are far more complex and varied than the satisfaction of economic security,[68] scholars, particularly of the political and social sciences, have argued that the educated middle class has largely formed the leading factions of environmental movements.[69] The fact that intellectual experts with specialised knowledge lead environmental organisations and movements in the West partly supports this argument. Despite contending ideas, we can hardly deny that the rising educated middle class was one of crucial forces behind the emergence of environmental awareness and the establishment of environmental organisations since the 1990s.

In South Korea, like-minded educated groups of people have generally been more environmentally conscious and concerned with environmental issues, too. With regard to the income level of the activist groups, however, it is not easy to apply the logic of middle class-educated-environmental awareness directly to South Korean society because it is challenging to grasp the entity of the local middle class; since the income gap widened at the end of the twentieth century, the middle class has been starkly divided into the upper and lower middle classes.[70] Therefore, in the South Korean situation, the logic of the middle class-educated-environmental awareness tended to mean that of the upper middle class-educated-environmental awareness.

Moreover, when it comes to the beneficiaries of environmental movements, we need to apply the more specified understanding of the middle class. In the South Korean context, the term 'middle class' is closely related to the term 'the public', and both are often used interchangeably. Therefore, by examining the problem with the term 'the public', we can inversely explore the trouble with the term 'middle class'. When the City or environmental organisations mention 'the public', it has a nuance of 'inclusiveness' or 'all'; yet, in reality, they both almost always focus on the middle class. As mentioned above, however, in South Korea, where the range of the middle-income class has been polarised since the 1990s, different organisations could strategically use the term 'middle class' in different situations, tacitly implying either the upper middle or the lower middle class for their own purposes.

A dispute between the City and environmental organisations surrounding the construction and operation of the golf course in Nanjido Post-Landfill Park demonstrates that they could not reconcile this issue because of their different interpretations and uses of the term 'the public' or the 'middle class'. The City thought 'the public' meant the upper middle class who could afford to play golf as a leisure activity both financially and psychologically, whereas the environmentalists believed 'the public' included the lower middle class who, they assumed, had psychological difficulty viewing golf as a leisure activity.[71] What is notable throughout this dispute is that the lower income class was entirely yet tacitly excluded. The limitations of the environmental movements of today lie in the way they confine their concerns for the physical environment to people of a specific income class.

The definition of the term 'the public' and its use are not only related to environmental movements but also to government-led urban planning in general because the government and any other organisation of power almost always exclude the lower income class when they mention 'the public', especially regarding environmentally conscious plans. In the urban redevelopment plans of the new century, the city and/or the developers claim that projects will better the lives of 'the public', but they often entail regional gentrification. The Sangam-dong New Town development was not an exception, since it gentrified the relatively underdeveloped town of the North Sangam-dong area by replacing the lower middle class or lower class indigenous inhabitants with new upper middle class residents. It even went so far as to transform the region into a commercial business district for large corporations. Since Nanjido Post-Landfill Park was intended to serve the public's wellbeing, the identity of 'the public' or the beneficiaries of the park's environmental and cultural values would be a matter of concern.[72]

Nanjido Post-Landfill Park for 'the public'

In order for Nanjido Post-Landfill Park to properly function as an inclusive public park, first, the leaders of the park-building—e.g. the city authority and landscape designers—need to have clear ideas about the members of 'the public' in relation to their income class and socio-economic situations so that they could determine their physical and psychological accessibility. Second, the Park's leisure programmes and required facilities have to be designed for public use as well as for the conservation of the natural environment for sustainable use.

Nanjido Post-Landfill Park is a landscape park designed and constructed for the public's in-city leisure activities, but it had to be geographically, financially and informationally accessible to function as an inclusive public park. Geographically, the Park is accessible within an hour using any public or private method of transportation from almost all areas of the city of Seoul, so its location is ideal. In reality, however, for an active leisure activity, such as camping, personal car-ownership is necessary. As a public park ('public' here means government-run) officially open to all for free—except for renting certain areas for special leisure activities or events—it is, in principle, financially accessible for all classes. According to Hyun-ji Kim's survey for her research on Seoul citizens' approach to leisure activities (2013), people of the higher income class (average monthly income over 5,000 USD) answered that being short of time prevented them from enjoying leisure activities, whereas those of the lower income class (average monthly income under 1,000 USD) answered that financial obstacles primarily kept them from leisure activities.[73] The financial matter, here, does not necessarily refer to the precise amount of money needed to access certain leisure activities in the Park, but to overall economic leeway; for example, although the minimum fee for the camping site per night is 10,000 to 13,000 KRW (10 to 13 USD), campers must bring their own camping equipment, so they also need to own a personal car.[74] Lastly, access to information ultimately determines the degree to which certain income classes can or cannot participate in the Park's leisure activities, but people in the lower income class were short of information on the leisure activities available to them.[75] In summary, although Nanjido Post-Landfill Park meets the conditions for the public's leisure activities in its structural design, its relatively more active leisure programmes exclude the lower income class for economic reasons.

Additionally, full access to the Park is decided by what people can do in the site—the programme. According to Hyun-ji Kim's survey, lower class people prefer relaxing and walking to activities that require equipment or skills.[76] Nanjido Post-Landfill Park is originally designed for multiple activities including both active practices like camping and sports and passive activities like walking and relaxing. However, more visitors to Nanjido Post-Landfill Park take part in the active leisure activities like camping than in the relatively passive ones. For example, despite the well-managed landscape, people only visit and use the top mounds of Haneul Park (Landfill 2) for special events or tourism rather than for relaxing or walking.[77] In effect, Nanjido Post-Landfill Park serves more purpose-oriented activities than less purposeful practices, which are typically considered the essential functions of the landscape park. In general, neither the Sangam-dong New Town residents nor people from other regions use the Nanjido Post-Landfill Park as an amenity for habitual reasons. Instead of making a journey to Nanjido Post-Landfill Park, people living in other towns would use amenities in their own neighbourhoods—this implies that in people's minds the post-landfill park is not equivalent to natural parks or mountains.

Considering that Nanjido Post-Landfill Park further serves people who have socio-economic leeway, at least beyond that of the lower middle class, the lower class was excluded from the concept of 'the public' in the Park design and development. This tendency to target the middle class is also related to the characteristics of the global style of parks, which caters to more affluent consumer groups that are able to exchange time and finances for the park's environmental and cultural, or symbolic values.

Nanjido Landfill's transformation into Nanjido Post-Landfill Park demonstrates new waste management techniques and trends in the new era of park-building during South Korea's socio-economic turn to the global neoliberal economy. Regarding waste management, the technical sealing, bordering and covering of the waste with a park of some sort is another way to control waste; making the unsightly invisible and burying the unsanitary, or the inappropriate, into oblivion. As a result, it reinforces the dichotomous division between waste and non-waste. In this sense, the Post-Landfill Park itself can be viewed as a part of the waste management system that annihilates the waste.

Built in the context of the environmentalism of today's global economy, Nanjido Post-Landfill Park has constantly produced environmental and cultural exchange values. On the one hand, the market economy easily co-opted these values, turning the Park's practical use value into symbolic value—for example, a park as a monumental structure, transformed from a landfill site, encourages park-tourism. On the other hand, these values, with their denotative positivity, tend to veil any negative aspects of the landfill's transformation into a park. Given its connection to the global economy, a park of contemporary environmentalism holds the risk of falling into an anthropocentric approach to the natural environment by emphasising the sustainable use of nature.

Nanjido Post-Landfill Park, which manifests itself as a public leisure site, harbours traits that are contradictory to the concept and true identity of 'the public'. While the term 'the public' ambiguously meant 'inclusive', the Park's design and programming focused on the middle class (inclusive of the lower and upper middle classes), and the Park's 'public' has almost entirely excluded the lower class. This exclusion suggests the subtle marginalisation of the lower income class, or the economically inappropriate. Without viewing the relationships between the natural environment, social dynamism and human subjectivity, Nanjido Post-Landfill Park may stay subordinated to the economic system and serve only a limited group of people.

Notes

1 The official Korean term *an-jeong-hwa* means 'stabilisation': cleaning and balancing land that has been contaminated with toxic materials. As it also means returning both the damaged natural and built environment back to its unpolluted state, I use the term 'regeneration' in tandem with stabilisation. Depending on the context, either term can be used independently.

2 In *The Art-Architecture Complex*, art historian Hal Foster calls the contemporary large-scale architecture, designed by renowned architects, which often become a spectacle of the urban space, the 'global styles' of architecture. In the sense that the landscape architects design the park under globalised neoliberal economic circumstances that incorporate and co-op the park into economic boundaries, and the park itself becomes a spectacle while producing the highest

benefit for the park and its region, I call this type of park the 'global style of park'. See Hal Foster, *The Art-Architecture Complex*, New York: Verso, 2011, Part 1.

3 In the 1970s, political scientists, including Allan Schnaiberg and Ronald Inglehart, described the transition from the pursuit of a quantity-based material life to that of a quality-based one as a turn to environmentalism. This study, however, interprets the term 'environmentalism in the context of global economy' as an ideology that uses any practices related to nature as a source of exchange value. See Allan Schnaiberg, 'Politics, Participation and Pollution: The "Environmental Movement"', *Cities in Change*, John Walton and Donald Cams eds., Boston: Allyn and Bacon, 1973: 605–627 and Walter Rosenbaum, *The Politics of Environmental Concern*, New York: Praeger, 1973.

4 See *Proceedings of Nanjido Landfill Stabilisation Construction* (2003a) and *Making of the World Cup Park* (2003b) published by Seoul City.

5 Although the government and environmental organisations used the terms 'public' and 'citizen' interchangeably, this study uses and discusses the term 'public' in tandem with the 'middle class' with regards to the park-building and its use, which is related to income class. As the term 'citizen' is concerned with nationality and citizenship as well as social class, it may digress from the discussion of this subject.

6 Robert Weller claims that the two phenomena were imported from the West to other regions, yet simultaneously, emphasises that globalisation cannot be viewed as a diffusion from a single core to the rest of the world. See Robert Weller, *Discovering Nature: Globalization and Environmental Culture in China and Taiwan*, Cambridge: Cambridge University Press, 2006.

7 Weller, 2006: 4–6.

8 Seoul City (Headquarter of Cleaning Projects), *A Preliminary Plan for Overground Landfill of Nanjido* (1985a).

9 'Nanjido Park Plan Too Many Risks', *Dong-A Ilbo* (16 October 1985). See also Seoul City (Department of Cleaning, Environmental Administration), *A Report on the Environmental Impact of the Overground Landfill of Nanjido* (1985b).

10 'Nanjido, A Rosy Dream from Waste Dump: Transforming into Park and Sports Complex in 10 Years', *Hankyoreh* (4 October 1988). Toxic and non-toxic industrial waste, including heavy metals (e.g. mercury, cadmium and used oil), was dumped in the unofficial industrial waste-dumping site in Landfill 1 (approximately 132 m²). As of 1988, 40 tonnes of industrial waste were dumped per day, reaching a total of 600,000 tonnes.

11 The housing development on the landfill site required the movement of all waste to other locations, such as Sudogwon Landfill (the new landfill in the Gimpo area). As of 1991, if 500 ten-tonne trucks delivered the waste daily, it would take 5 years and a budget of 4–5 billion USD to move the total amount of waste of approximately 70,000,000 tonnes. See 'Suggest to Develop Nanjido into Housing Complex', *Dong-A Ilbo* (23 October 1991). Just before the landfill's stabilisation and regeneration construction began, the conservative party insisted on the removal of the mass of waste to develop a new town on the flat land. See 'Seoul City: Environmental Park, High-tech IT District vs. Minja Party: Housing Development', *Hankyoreh* (1 June 1993) and 'Suggest to Build Administration Town in Nanjido', *Kyunghyang Shinmun* (15 October 1996).

12 Seoul City (with Samsung Construction Co. Ltd.), *A Study on the High-density Business Town Development of Seoul* (1992c); 'Development Corporations Licking their Lips at the Nanjido Region: Seoul City's Park-building vs. Samsung's Teleport', *Hankyoreh* (9 November 1992).

13 As soon as the 2002 FIFA World Cup main stadium site was announced, Sangam-dong emerged as a hot real-estate area for investment. See 'Grab Jamshil and Sangam-dong New Town', *Kyunghyang Shinmun* (9 November 1998).

14 In 1999, Seoul City held an international symposium for the development of Nanjido Post-Landfill Park. The City invited overseas landscape architects/planners who were involved with projects dedicated to transforming closed landfills into a park.

15 Paul Thompson, 'Building on Marginal and Derelict Land: Stockley Park London', *Envisioning the Millennium Park*, Proceedings of the International Symposium towards the Sustainable Development of Nanjido (The Seoul Institute and University of Seoul: 1999): 1–2.

16 The Urban Land Institute (ULI) highly regarded the landscape design of Stockley Park for the way it had reclaimed a former garbage dump and the way its technology-oriented business park set the trend (ULI's development case studies, 2001), showing the possibility of combining the Business Park and public leisure space. Nevertheless, the Park has been known for its dreary atmosphere as it is nearly empty of traffic and has retained its status as a hidden landscape park in London.

17 On Arup Associates' Stockley Park phase 3 plan, see www.arup.com/projects/stockley-park (accessed on 20 April 2018).

18 Thompson, 1999: 4. The successful construction of a golf course and its positive impact on the natural environment in Stockley Park provided the foundation for Seoul City's insistence on building the golf course on Nanjido Post-Landfill Park.

19 Mary Margaret Jones (a senior principal at Hargreaves Associates), 'Urban Parks by Hargreaves Associates', *Envisioning the Millennium Park*, 1999: 9–10.

20 http://ladprofile.weebly.com/george-hargreaves-1952.html and www.peterrichardsart.com/byxbee-park.html (accessed on 20 April 2018). Many of the poles no longer stand vertical because of the subsidence of the garbage buried underneath. In 2013, as the City of Palo Alto decided to resume the development over the whole Park area, it dismantled many works of art in the Park. See Brad McKee, 'The Dismemberment of Byxbee Park', *Landscape Architecture Magazine*, American Society of Landscape Architects (29 October 2013).

21 Jo Moss, 'Redeveloping Homebush Bay', *Envisioning the Millennium Park*, 1999: 1.

22 For the debate on the environmental problems and suppression of public discussion of the toxic waste, see Sharon Beder, 'Sydney's Toxic Green Olympics', *Current Affairs Bulletin*, Vol. 70, No. 6, November 1993: 12–18.

23 The operation of Freshkills Park has now been handed over to the NYC Department of Parks and Recreation and the NYC Department of Sanitation. For information on the building processes and connotations of Freshkills Park, see Mohsen Mostafavi and Ciro Najle eds., 2003; Charles Waldheim ed., 2006; Caroline Klein et al. eds., 2013; and James Corner, 2014.

24 The processes and short-term evaluation of the Nanjido Landfill's stabilisation and reclamation draw on the bio-chemical and geological research conducted by the Seoul Institute in 2000. The Seoul Institute, formerly the Seoul Development Institute established in 1992 with approximately 250 researchers, is the major research centre for Seoul's short-term and long-term urban development policies.

25 Researchers from the Seoul Institute predicted that the stabilisation and reclamation, without waste with a slow rate of decomposition, would reach 70% by 2000 and 99.7% by 2029 (The Seoul Institute, 2000: 47–48).

26 Ibid.: 77–81. Throughout the 1990s, the rate of Ammoniacal nitrogen far exceeded the standard level, while the rate of most other chemicals was reduced under the standard of permission.

27 Ibid.: 107, 110.

28 Nanjido Landfill, immediately after its closure, consisted of industrial sludge (0.7%), municipal solid waste (23.8%), used construction materials (16.8%), soil (22.2%) and non-drilling parts (consisting of inorganic substances) (36.5%) (Ibid.: 51–55).

29 Ibid.: 96–99.
30 The US National Research Council (NRC) defines that risk assessment is to assess the potential health impacts caused by bodily exposure to environmentally hazardous elements. The US National Academy of Sciences suggested four steps to assess risks: 1) Hazard Identification; 2) Exposure Assessment; 3) Toxic Assessment; and 4) Risk Characterisation. The Seoul Institute proposed the use of this method to monitor the air pollution in the Nanjido post-landfill site (Ibid.: 169; www.epa.gov/fera/nrc-risk-assessment-paradigm. accessed on 20 June 2016).
31 '[To resolve the odour problem] the landfill stabilisation, including the gas collection and treatment, had to be completed by the first half of 2001 in one way or another in preparation for the FIFA's on-site inspection in November 2001' (Ibid.: 139, 141).
32 As Seoul City planned to change the use of the land from 2003, they stopped farming in early 2002.
33 As for deodorising techniques, they selectively applied biological and chemical methods (e.g. biofiltering and activating charcoal absorbents and neutralisers) depending on the cause of the odour, deodorising effectiveness and economic feasibility. Regarding the control of the odour from the Nanji sewage treatment facility, the researchers proposed covering the facility or spraying deodoriser during the World Cup games—the City chose the latter (The Seoul Institute, 2000: 122–155, 137).
34 Ibid.: 147.
35 Ibid.: 149–150.
36 The City and the citizens alike considered the leachate treatment facility a detestable structure because it generated odour during its multiple processes of leachate treatment. They, thus, located it far from residential and commercial areas to prevent any physical or psychological factors that may negatively influence the development of the Sangam-dong New Town and inbound tourism (Ibid.: 154).
37 Ibid.: 204–239.
38 The expression 'eco-park' is widely used for parks built based on a concern for natural environment.
39 The Seoul Institute, 2000: 222.
40 Weller, 2006: 5. For more on this, see Keith Thomas, *Man and the Natural World*, London: Penguin, 1991: 242–287 and Simon Schama, *Landscape and Memory*, New York: A. A. Knopf, 1995.
41 Regarding Frederick Law Olmsted and the building of Central Park in New York City, see Roy Rosenzweig and Elizabeth Blackmar, *The Park and the People: The History of Central Park*, Ithaca, NY: Cornell University Press, 1998. Weller also mentions the park creations of the United States during the late nineteenth and early twentieth centuries along with the global park-buildings of today (Weller, 2006: 69).
42 Weller, 2006: 72–73. For more on the idea of nature in relation to transcendentalism and pragmatism, see Woodbridge Riley, 'Transcendentalism and Pragmatism: A Comparative Study', *The Journal of Philosophy, Psychology and Scientific Methods*, Vol. 6, No. 10, May 1909: 263–266 and Mark Mumford, 'Form Follows Nature: The Origins of American Organic Architecture', *Journal of Architectural Education*, Vol. 42, No. 3, Spring, 1989. See also R. W. Emerson, 'First Visit to England' in *The Complete Works of Emerson Vol. V* (English Traits), Ch. 1; 'Plato; or, the Philosopher' in *The Complete Works of Emerson Vol. IV* (Representative Men), Ch. 2; and 'Art' in *The Complete Works of Emerson Vol. VII* (Society and Solitude), Ch. 3 along with Emerson, *Nature*, Boston: James Munroe and Company, 1836.

43 It is worth understanding how differently Asian and Anglo-American cultures—
especially the US culture, which has heavily influenced South Korean culture
in general—have approached nature, which helps us examine the subtle differences between countries under global environmentalism. Natural history in the
United States, particularly in New England, is grounded on transcendental ideas
about the relationship between God, nature and human beings, in which nature
is regarded as the presence of God. The National Audubon Society, a non-profit
environmental organisation dedicated to conservation, was concerned with bird
watching (first established in Massachusetts in 1897, which was incorporated
with societies established in other states in 1905). Precursors in the UK include
movements to protect species that were being decimated by the feather trade. See
Weller, 2006: 69; Felton Gibbons and Deborah Strom, *Neighbors to the Birds: A
History of Birdwatching in America*, New York: W. W. Norton and Company,
1988. See also Royal Society for the Protection of Birds, *History of the RSPB*.
2004 (18 March 2004). www.rspb.org.uk/about/history/index.asp.

44 In this part, I use the term 'environmentalism' to indicate the early stage of environmental consciousness and responsible approaches without the critical connotation of its co-option with the capitalist economy.

45 Joe Harry and Joseph Harry, 'Causes of Contemporary Environmentalism',
Humboldt Journal of Social Relations, Vol. 2, No. 1, Fall/Winter 1974: 3–7;
Schnaiberg, 1973; and Rosenbaum 1973.

46 Other scientists-activists include Henry Fairfield Osborn, William Vogt, Aldo
Leopold, Barry Commerner and Paul Ehrlich (Joe Harry and Joseph Harry,
1974: 4).

47 Weller, 2006: 76–77. See also Nigel Dudley ed., *Guidelines for Applying
Protected Area Management Categories*, Gland, Switzerland: IUCN, 2008,
2013. Areas protected under category 4 are generally publicly accessible (19);
those under category 5 aim to produce an area of distinct character through
interactions between people and nature (20); and areas under category 6 are
fundamentally intended for human use (22).

48 Joe Harry and Joseph Harry, 1974: 5. The authors view all three phenomena on
the same line as the causes of environmentalism of the times.

49 The education includes the re-use of methane gas as power sources for neighbouring areas and the reclamation of recyclable raw materials to make products
such as bricks and so forth.

50 In 2008, the original 18-hole golf course ended operations after a dispute
between Seoul City and the Korea Sports Promotion Foundation (2004–2008).
In 2014, the City changed the golf course into a reduced-sized casual park golf
course, using the land from one hole of the original course.

51 These are more group-oriented leisure activities than individual endeavours.
On the study of American parks, Roger Kennedy, director of the National Park
Service during the Clinton administration, stated that immigrants from Africa,
Southern Europe, Southeast Asia and Latin America have strong traditions
of family and clan gatherings (Setha Low, Dana Taplin and Suzanne Scheld,
Rethinking Urban Parks: Public Space and Cultural Diversity, TX: University
of Texas Press, 2005: 42; J. Woolf, 'In Defense of the Metropolitan Mosaic',
National Parks 70, Jan–Feb 1996: 41). Although the American and South
Korean cases have different group compositions, this study implies that there
are significant tendencies for group-oriented collective activities amongst Asian
populations.

52 In explaining the Japanese style of parks, Weller points out that the parks in
Japan rarely allow leisure activities except for forest bathing, boot camp programmes, group tourism and golf playing (Weller, 2006: 74–76). However,

we cannot determine if golf playing and group tourism are from Japanese-style parks because the two activities are the most significant forms of leisure activities in the global style of parks, too.

53 Nanji Han River Park is located by the Han River. It is one of the five sub-parks of Nanjido Post-Landfill Park and also one of many Han River parks, which have riverside bike lanes and allow water play. The management office of this park is different from that of the main Nanjido Post-Landfill Park.

54 Weller, 2006: 69. Weller's term 'natural park' refers to a park that is built based on and aims for the preservation and appreciation of nature, which I interpret as the same as the 'landscape park'.

55 In the postmaterialist discourse, Ronald Inglehart, along with other scholars, used the term 'environmentalism' to refer to responsible approaches to the environment in general. When discussing the postmaterialist discourse, I use the term 'environmentalism' as the scholars had in their discursive context. I use this term because there were various critical opinions about the environmentalism of that time, particularly because of its human-centred standpoint.

56 Postmaterialism was one of the topics of these scholars. Ronald Inglehart coined the term in *The Silent Revolution: Changing Values and Political Styles Among Western Publics* (NJ: Princeton University Press, 2015 [1977]). The primary hypothesis of postmaterialism is that those who have experienced sustained high material affluence give priority to values such as individual improvement, personal freedom, the ideal of a society based on humanism and the preservation of a clean and healthy environment.

57 Peter Dauvergne, 'The Illusions of Environmentalism', *The Shadows of Consumption: Consequences for the Global Environment*, Cambridge, MA: MIT Press, 2008.

58 Riley Dunlap and Angela Mertig, 'Global Environmental Concern: An Anomaly for Postmaterialism', *Social Science Quarterly*, Vol. 78, No. 1, March 1997: 24–29.

59 Sarah Ray, 'Toward an Inclusive Environmentalism', *The Ecological Other: Environmental Exclusion in American Culture*, AZ: University of Arizona Press, 2013: 179–184.

60 Alexander O'Neill, 'What Globalization Means for Ecotourism: Managing Globalization's impacts on Ecotourism in Developing Countries', *Indiana Journal of Global Legal Studies*, Vol. 9, No. 2, Spring 2002: 501–528 and Donald G. Reid, 'Globalization and the Political Economy of Tourism Development', *Tourism, Globalization and Development: Responsible Tourism Planning*, London: Pluto Press, 2003.

61 Gilbert Rozman, 'The Northeast Asian Regional Context for Environmentalism: Assessing Environmental Goals against Other Priorities in the 1990s', *Journal of East Asian Studies*, Vol. 1, No. 2, Special Issue: Perspectives on Environmental Protection in Northeast Asia, August 2001: 13–30.

62 Weller takes the position that globalisation, by arguing that there is no single core, is a creation of a broad and direct diffusion of ideas through various colonial and postcolonial mechanisms of transmission to the rest of the world. Drawing on the thoughts of Immanuel Wallerstein and Walter Rostow, he also refers to the early argument about whether globalisation is a set of independent reactions to shared economic problems or the result of direct power relations in a single system. See Weller, 2006: 6; Walter Rostow, *The Stages of Economic Growth: A Non-Communist Manifesto*, Cambridge: Cambridge University Press, 1960; and Immanuel Wallerstein, *The Modern World-System: Capitalist Agriculture and the Origins of the European World-Economy in the Sixteenth Century*, CA: University of California Press, 2011 (1974).

63 It has been conceived that the United States played a leading role within the IMF, controlling the international transfer of capital. In the discussion of globalisation, Fredric Jameson commented that the IMF has been the driving force of neoliberalisation by imposing free-market conditions on countries in financial trouble by threatening to withdraw investment funds (Fredric Jameson, 'Globalization and Political Strategy', *New Left Review* 4, Jul–Aug 2000: 56).

64 Regarding the shift of South Korea's political economic system from developmentalism to neoliberalism, including internal and external influences and socio-political particularities, see political economist Joo Hyoung Ji, 'Learning from Crisis: Political Economy, Spatio-Temporality, and Crisis Management in South Korea, 1961–2002', Dissertation at Lancaster University, 2005. According to Joo Hyoung Ji's research, when a democratic administration officially took office in 1987, neoliberalists, including a group of economic specialists educated and trained in America, began to establish the preconditions for neoliberal economic systems, such as market-opening, deregulation and employment flexibility.

65 The Federation inherited the nation's first environmental organisation called the Korea Centre for Pollution Research, which was established in 1982.

66 Weller, 2006: 5.

67 That the environmental activists were not distressed by the relocation of the Nanjido Landfill population at the time of the landfill's closure—which was closely related to the inhabitants' employment and sustainable socio-economic living situation—because the residents received a reasonable amount of financial compensation also demonstrates that environmental movements in South Korea were and are limited to the conservation of the natural environment, heedless of the correlations between matters of the environmental ecology and social ecology.

68 Steven Brechin and Willett Kempton, 'Global Environmentalism: A Challenge to the Postmaterialism Thesis?' *Social Science Quarterly*, Vol. 75, No. 2, June 1994: 247.

69 Inglehart was at the forefront of the argument on the relationship between middle class education and environmental awareness. See Ronald Inglehart, *Culture Shift in Advanced Industrial Society*, NJ: Princeton University Press, 1990. Political scientists Quentin Kidd and Aie-Rie Lee defend Inglehart's argument against contending opinions (Quentin Kidd and Aie-Rie Lee, 'Postmaterialist Values and the Environment: A critique and Reappraisal', *Social Science Quarterly*, Vol. 78, No. 1, March 1997: 1–15).

70 According to the statistics of the income and class differences in 2016 South Korea, the class gap has grown wider each year (income quintiles increased from 7.59 in 2013 to 8.08 in 2014, 8.24 in 2015 and 9.32 in 2016): for example, based on the 2016 statistics, if the lowest 20% earn 1,000,000 KRW per month (1,000 USD), the top 20% earn 9,320,000 KRW per month (9,320 USD). Also, the proportion of the mid-income class decreased, while the low income class increased (mid-income: low income—65.6 : 14.6 in 2013, 65.4 : 14.4 in 2014, 67.4 : 13.8 in 2015, 65.7 : 14.7 in 2016). From Statistics Korea http://kostat.go.kr/portal/korea/index.action (accessed on 26 June 2017).

71 The golf course did not close because of the environmental organisations' triumph over the government but because of the financial issues between Seoul City and the Korea Sports Promotion Foundation.

72 Nanjido Post-Landfill Park did not replace the landfill residents with other groups of people, but annihilated and buried them in oblivion; this was not only for their income class level (some earned an amount equivalent to that of the upper middle class), but also for the uncleanliness of their occupation, or for the way it was inappropriate to the social norm.

73 Hyun-ji Kim, 'Perception Gap between Social Strata of Leisure Activity Diversification: Focusing on the Han River Park in Seoul', Master's thesis at University of Seoul, 2013: 91.

74 For an overnight stay for a family of four in the campsite, the rental fees for basic equipment (e.g. a tent, sleeping bags, blankets, etc.), without food, is from approximately 100,000 KRW (100 USD) to 140,000 KRW (140 USD). For the lower income class with its monthly income of 1,000,000 KRW (1,000 USD), an overnight camping experience costs nearly one fifth of their monthly income.

75 Hyun-ji Kim, 2013: 87, 102.

76 Ibid.: 89, 93.

77 The top of Haneul Park (Landfill 2) was initially intended to be a themed eco-park for relaxing and walking as well as for learning about the ecosystem. Because of its height, however, the park is used for special events or tourists' visits. Tourists from other provinces and from abroad visit the park for educational purposes; they look around its historical makeover, regarding the site as a monument for environmental regeneration, rather than appreciating the nature in the park. Overall, Nanjido Post-Landfill Park's symbolic meaning overshadows its practical functions.

5   Art

Disruption of Nanjido Post-Landfill Park

Positioned atop closed landfill mounds, Nanjido Post-Landfill Park creates an unusual sense of place or placelessness. This sensation is rarely experienced in other parks that function as amenities for the community's well-being. This chapter begins with a discussion on the sense of unease and its relation to place or placelessness. Then, it attempts to demonstrate the sense of unease or placelessness perceived in Nanjido Post-Landfill Park. It also shows how artists have engaged with those subjects by considering the site-specific art projects produced by the SeMA Nanji Residency artists, focusing on the works by Wonho Lee and Joon Kim. Prior to the analyses of the two art projects, I will examine other forms of art that deal with waste and land-fills (e.g. through photography and performative engagement). Looking at their characteristics will help us to understand how artists have made or salvaged the value of the 'wasted', which is associated with the sense of place or placelessness. These discussions of creativity will enable us to examine the invisible and ungraspable sense of place or placelessness, a quality that other disciplinary methods may not be able to describe. In this context, the aesthetic approach creatively undertakes a critical commentary on the human reality, which is grounded on truthful and ethical attitudes. In this way, I argue that material and social sanitisation has defined the urban landscape of contemporary Seoul and influenced the production and elimination of the sense of certain places.

Waste, as a part of the value system, refers to over-production—or in economic terms, the surplus—that is deemed valueless by socio-economic criteria and, as a result, outcasted from the social realm. Here, waste includes not only wasted materials but also wasted humans (e.g. landfill garbage collectors whom the society often likens to the waste with which they associate). The value of material waste is inextricably intertwined with the value of the people, their professions and the environmental and socio-economic living conditions. To explore the complex value system of materials and human beings (as Bauman calls, the surplus population[1]), artists, in the tradition of politically aware conceptual art, emerged in the 1960s and 1970s, prompting a commentary on the human reality.[2] These artists employ aesthetic methods to re-evaluate the values of the abandoned, and, going beyond, re-present or re-embody the abandoned material and their existences. Since their work re-categorises the wasted, it is a disruptive practice that addresses the interdependent feedback system of the ecology.

The most essential approach taken to salvage the value of surplus objects and the surplus population is to reclaim the abandoned or hidden value of the material and/or humans and reconstitute them in the existential world where materials endure almost permanently. Although they may exist in different forms, they are re-used in cyclical ways. One of the most prevalent forms of the art of reclamation is a sculptural installation created from recyclable material sources, re-making something so that it is useful and meaningful.[3] Artists have also revived the value of the human subject through performative practice, a direct involvement or participation with the socio-political issues of human beings and their environment. They accomplish this by engaging with individuals from a community that has been abandoned or wasted for material or social sanitary reasons.[4] Striving to rediscover the value of the wasted, this art form is, however, only possible in the current unsanitary landfill sites where human hands still manually sort recyclable materials.

In the closed landfill sites, as in the case of Nanjido Post-Landfill Park, reclaiming the value of wasted materials (in the process of regeneration) and humans (the previous residents of the Landfill site) requires aesthetic approaches. On the one hand, the artists' search for evidence, which would enable them to make connections between the past and the present, or conflated layers of histories, is grounded on material excavation. On the other hand, it is an aesthetic endeavour as the artists' work considerably depends on their poetic imaginaries of the spatio-temporal dimensions of the site.[5] In other words, their process seeks what was once here or what is still here but suppressed, hidden or invisible, and perceived as non-existent consequently. Their efforts then make the site a place. In summary, an aesthetic practice based on material excavations looks for something that is no longer visually and tangibly extant, but has a trace of existence that the artist can excavate or at least capture through non-visual senses, such as the olfactory and the auditory. Their process is inverted because it can reveal the cause of the site's placelessness, if not revive the place itself, by unveiling the reality of the materials' and humans' invisibility and intangibility and the absence of the place.

The artworks of the SeMA Nanji Residency artists, which I analyse in this study, do not particularly invite the public's participation in the works of art, so it would be overreaching to discuss these pieces in the context of participatory public art. Nevertheless, these site-specific artworks undoubtedly question the relationship between art and the social by engaging with a part of the urban phenomena that is, in itself, social and political. While these works do not explicitly demonstrate or challenge socio-political issues in the form of public art, they do address the ways in which artistic practices constitute and enhance social experiences and in which social experiences trigger artistic practices. Claire Bishop, in *Artificial Hells* (2012) cites Guattari's notion of transversality in this context of art and the social:

> [Transversality] offers one such way of thinking through these artistic operations: he leaves art as a category in its place, but insists upon its constant flight into and across other disciplines, putting both art and the social into question, even while simultaneously reaffirming art as a universe of value.[6]

She then asserts that we need to transform the existing institutions through the transversal encroachment of bold ideas that are related to the artistic imagination.[7] Likewise, the purpose of artistic projects, especially the site-specific works on Nanjido Post-Landfill Park, is transversally to smear the artists' bold ideas into the frame of existing dichotomous thoughts on the site (past-landfill-dirty vs. present-park-clean).

First, this chapter will use aesthetic analyses to link the material with the metaphoric sense of social waste to clarify how the idea of material waste traverses and helps us to understand social waste comprehensively. Second, the analysis of the aesthetic apparatuses that artists applied to reveal the site's sense of place or placelessness will explain how the transformations of the Nanjido region—from wetlands to a landfill to a park—demonstrate the shifting concepts of waste in association with the changing socio-historical circumstances and varied approaches to the natural environment. These discussions of aestheticism will help us to address the importance of interconnectedness that must be considered in the study of urban ecology.

Unease and placelessness

The sense of unease

The SeMA Nanji Residency artists' site-specific art practices on Nanjido Post-Landfill Park, particularly those of Wonho Lee and Joon Kim, proceeded from their curiosity about the sense of unease of the current Park space. On the subject of place, the sense of unease of a place does not necessarily mean that the site is not a place. Nevertheless, we cannot completely deny that there is a relationship between the two concepts, so a closer examination is required to determine if Nanjido Post-Landfill Park's sense of unease is related to the site's sense of place or placelessness.

The terms 'uneasiness' and 'unease' are often used interchangeably. By definition, uneasiness pertains to the symptoms of an uneasy state: for example, the state of 'anxiety', 'discomfort' (Oxford dictionary), 'worry' (Cambridge dictionary), 'the restless', 'the disturbed', 'the perturbed' and 'not being easy in body or mind' (dictionary.com). In gender discourse, this word is used in relation to the matter of cross-gender, an identity that cannot be pinned down to one of the two sexualities. In the discussion of geopolitics, it is used to indicate a situation that is uncomfortably located between two sides.[8] The term 'uneasy', the root of the word 'uneasiness', is defined as 'feeling or showing uncomfortable feelings of uncertainty' (Merriam-Webster dictionary). Meanwhile, it is notable that the term 'unease' is defined as 'an uneasy state of mind usually over the possibility of an anticipated misfortune or trouble' (Merriam-Webster dictionary). Although the term 'unease' specifies the possibility of an anticipated misfortune or trouble, I focus on the meaning of the unrealised, thus, uncertain potential state, which is actually open to possibility. Therefore, I argue that the two terms, uneasiness and unease, share the symptomatic senses, so I consider these words in the same line of thought, focusing on the feelings that stem from uncertainty and the unknown. In the discussion of the sense of place or placelessness of Nanjido Post-Landfill Park, however, I use the term 'unease' on the premise that visitors or users of the Post-Landfill Park are mostly aware of the fact that beneath the Park mounds remains the former Landfill, which may or may not be potentially toxic.

Place and placelessness

A look into the history of the discussion on 'space', 'place', 'sense of place' and 'placelessness' will help define the concepts of 'place' and 'placelessness'. In the late 1970s, geographers Yi-Fu Tuan and Edward Relph took the lead in the studies of space and place, distinguishing one from the other. Tuan saw place as a pin-downed location within a space that allows constant movement and flow.[9] Drawing on Tuan's idea, Liz Taylor views place as 'a meaningful location', or portion of space that combines the natural with the built.[10] Regarding the 'sense of place', Edward Relph states that physical characteristics and activities that take place in a location form its identity. He holds that these characteristics and the identity of a place constitute the 'sense of place', and interprets it as equivalent to the 'spirit of place', or 'genius of place' (*genius loci*).

> This is the attribute of identity that has been variously termed 'spirit of place', 'sense of place' or 'genius of place' (*genius loci*)—all terms that refer to character or personality. Obviously, the spirit of a place involves topography and appearance, economic functions and social activities, and particular significance deriving from *past events and present situations*—but it differs from the simple summation of these.[11] (Italics added)

For Relph, the social and economic activities that take place in the location (which is embedded in and embodies the historical times) as well as the topographical activities (which embodies the spatial frame of the place) mainly determine the sense of place. His acknowledgment of the site's significant past and present circumstances, or history, that constitute its sense of place emphasises the temporal aspect of place. However, it is not the individual activities or incidents, but the inter-woven web of topographical, socio-economic aspects that genuinely create the sense of place. Arguing that every place has its own sense of place and authenticity based on these geo-social attributes, Relph regards homogenised global cities as fake places, or sites of placelessness, for their lack of original sense, spirituality and genius.

On the other hand, since the late 1990s, Doreen Massey has asserted that 'a global sense of place' is the more relational constitution, stating that 'place' can be seen as a particular, unique point of intersection rather than its own meaningful location. Instead of areas surrounded by boundaries, Massey claims to view a place as a location where articulated movements transpire in the networks of social relations. She then suggests that a global sense of place can be possible by seeing it as a part of the relational networks of places.[12] She also argues that the uniqueness of a place remains valid and valuable within the structural changes of globalisation, which is in contrast to Relph's viewpoint on placelessness or the loss of a sense of place in the global era. Further expanding on her stance, Massey concludes that gaining insight into a place is only possible by linking it to other places.[13] By placing an emphasis on the social relations between places, Massey tries to unearth meaning from places of movements in the global era, but this does not mean that she completely denies the historicity of a site as one of the constituents of a place.

Marc Augé's conception of place is based on relations and flow; in *Non-Places* (first published in 1995), however, he demonstrates the matter of placelessness found in sites of circulation and constant mobility, such as airports and hypermarkets, through their lack of attachment to a traditional sense of location. In 2008 (in his second edition of *Non-Places*), he explores the problem of discontinuity and interdict in global cities, while further elaborating on the characteristics of placelessness or the sense of non-place. In his new edition published in the new millennium, Augé points out the more complicated conception of placelessness and non-place as a further fragmented utopia, divided between anxiety and hope.[14] Augé analysed large-scale urban architecture, drawing an analogy between the contemporary edifices and ruins, and proposed the idea that the former reproduces the relation with time expressed by the spectacle of the latter. He states:

> What we perceive in ruins is the impossibility of imagining completely what they would have represented to those who saw them before they crumbled. They speak not of history but of time, *pure time* [...] To perceive *pure time* is to grasp in the present a *lack* that structures the present moment by orienting it towards the past or the future.[15] (Italics added)

In this case, the past and the future may not be the historical moments of time, but imaginary spaces into which the lack of a present moment slips—the overall experience of historical moments in such a place is an illusion or spectacle. In the same vein, his research defines non-place as the place that:

> provides an experience—without real historical precedent—of solitary individuality combined with non-human mediation (all it takes is a notice or a screen) between the individual and the public authority.[16]

In his definition of place in 1995, Augé emphasises the attachment to a traditional sense of location, while in 2008, he further considers the perceptual elements of the human experience of a site. By elaborating on the way in which temporal lineage is perceived through the idea of *lack* in the global era's spectacularised sites, he introduced the evolved concept of imagined place through placelessness. In the overall discussion of place and placelessness, Augé has held on to the idea that place is somehow related to the site's historical times. Augé and Relph similarly insist on diachronic continuity as an essential condition for place. I regard both temporal and spatial relations as conditions for place.

When discussing social and economic activities as one of the essential conditions of place, we must clarify that the absence of activity does not give a site the quality of placelessness. Likewise, we cannot confirm that Nanjido Post-Landfill Park is a place simply because leisure activities take place in the Park. The issue, here, is that these activities do not accumulate to produce a historically meaningful time-space even though they may take place repeatedly and persistently. Instead, they are confined to their own momentary frame of time and point of space, disconnected from the site's historical (temporal) and regional (spatial) relations.

First, as for the historical discontinuity in Nanjido Post-Landfill Park, it is not a simple disconnection between the past and the present. Its past is indeed present but suppressed and unseen. This attribute causes the visitors and users of the site to oscillate between the sense of place and placelessness of the Park. In this sense, we can mark the *lack (-less)* of the sense of place of Nanjido Post-Landfill Park in brackets—place[less]ness. In other words, what surrounds Nanjido Post-Landfill Park is not simply the placelessness produced by the built environment itself, but the placelessness created by the social authority's psychology of abandonment or denial and the suppression of the site's history, which are responsible for the vertical axis of the place. The uncertainty about the past-present historical layers that underlie this site causes a sense of unease, too.

Second, the placelessness of Nanjido Post-Landfill Park within the urban context stems from its spatial disconnection from neighbouring regions. While Nanjido Landfill's transformation into Nanjido Post-Landfill Park made the material components of the previous era invisible, it emphasised the meaning of the Park in the new era as a symbolic site of the environmental and cultural values that are presumed to contribute to the public's wellbeing. In other words, Nanjido Post-Landfill Park's monumental characteristics separate it from the neighbouring urban space. Mainly visited by tourists for tourism and special events (including seasonal botanical events and camping), the Park is disconnected from the everyday living patterns of other neighbourhoods.[17] The City's emphasis on the 'ecological' aspect (which here means a responsible approach to the natural environment) of Nanjido Post-Landfill Park, as a result, Disneyfied and museumised the site[18]—it had become a themed space outside the urban context.

Artistic engagement with the urban space

Documentary photographs on the landfill

Photography is the most prevalently used medium for the subject of the landfill or the wasteland. Documentary photography, in particular, is useful for its journalistic and archival aspects. First, the journalistic or reportage aspect of documentary is effective in conveying the environmental and social concerns related to the landfill. Second, documentary projects are meaningful for their archival purposes; they present how the site has been established, transformed, remembered and represented, which constitutes an 'archive' of the city space.

Journalists or journalistic artists have recorded the conditions of current landfills—inhabited landfills or uninhabited landfills where waste workers do not reside but labour for waste management—in a documentary format through photography. They make records of the landfill environments, particularly those in developing nations, such as China, the Philippines, Cambodia, Brazil, Egypt and so forth.[19] These works often describe industrialisation's deterioration of the environment and the lives of the landfill habitants, which are interconnected with each other. The image-based journalistic approach, however, has a pitfall, as it tends to produce stereotyped imageries of the landfill environment and its people by representing them as victims at environmental risk from a philanthropic point of view. Even if the authors intend to represent the reality of the environmental problems facing humanity and the existential dignity of the landfill inhabitants, in some cases, viewers misinterpret these works as reports on an alien form of life in a distant part of the planet.

During Nanjido's landfill period from the 1980s to the 1990s, Korean newspapers and religious magazines occasionally reported on the Nanjido Landfill garbage collectors' community, lives and work, but they described the Landfill area and its people as disembodied entities. For instance, even though the Nanjido region is geographically located fairly close to the heart of the city of Seoul, reports illustrated the Landfill and its people as a socio-economically marginalised community, situated outside the imaginary boundary of the urban space of Seoul—it is genuinely difficult to find a news report that indicates the specific location of Nanjido Landfill. The

Seoul City government also took photographs of Nanjido Landfill regularly during the landfill period for archival purposes. It similarly produced stereotyped images of Nanjido Landfill and its inhabitants as the unusual dark side of the modern city of Seoul by focusing on the massive dumpsite, the toiling garbage collectors covered with dust and the residents' shabby housing conditions. In effect, the journalists' and government's archival photographs not only recorded Nanjido's landfill period but also contributed to creating and reinforcing stereotypes of Nanjido Landfill and its population (Figure 5.1). Journalists and Seoul City alike continued to take archival photographs of the Nanjido region during the Post-Landfill Park period, too (Figure 5.2). Seoul City, in particular, regularly photographs the site, documenting the landscape and the public's use of the Park. The Post-Landfill Park photographs, notably those taken by the City, focus on the site's

Figure 5.1 Nanjido Landfill. c. late 1980s. © National Archives of Korea

Figure 5.2 Noeul Park (Landfill 1) of Nanjido Post-Landfill Park. 2000s. Courtesy of Seoul City

successful transformation since they were taken for promotional purposes. Consequently, the promotional activities produced yet another stereotype of the Nanjido region of the new era: the rehabilitated natural environment.

Meanwhile, as Nanjido Post-Landfill Park was completed, photographs of the pre-landfill period emerged to emphasise that the landfill's transformation into a park embodies the restoration of the region's pre-landfill natural environment. So far, only the photographs of Jong-Chul Won, a former resident of the Nanjido region during the 1960s, have been presented to the public. Won's photographs depicted everyday scenes in the Nanjido region before the site turned into the municipal landfill (Figure 1.3). He portrays the pre-landfill Nanjido region as an unpolluted landscape where the people made their livelihood on farming. Won's photographs pay particular attention to the stream that made the area an isolated wetland, and the people are depicted as enjoying a bucolic form of life. Since Nanjido Post-Landfill Park's opening, his landscape scenes have been seen as the ideal for the natural environment's regeneration, and those of the people's lives, despite the practical hardships in their living conditions, have been thought to conjure nostalgic aspirations.

These photographs represent the Nanjido region of different periods. Non-Nanjido residents accept the history of the Nanjido area as presented in these photographic images, as the archival photographs lack a 'critical view on the reality' of the situation. The resurgence of the photographic scenes of the pre-landfill period also signifies that Nanjido Landfill's regeneration as a remediation of a degraded biological ecosystem used the natural environment of the pre-landfill period as a reference point. During this process, both the natural and social environments of the pre-landfill Nanjido region have become mythologised.

On the one hand, photography is one of the most effective media used to report, record and represent the urban conditions and transformations of a site over time. On the other hand, it also runs the risk of stereotyping the photographed environment and people—in extreme cases, making them into spectacles thwarting the viewers' access to the site's other invisible or under-represented spatio-temporal layers, thereby dismissing its accumulated history and relations with other regions that embody the place. The works of art I analyse in this study focus more attention on material explorations of the historical sequence than on the spatial imageries.

Place and memory-image

In his discussion on photography and history, Siegfried Kracauer argues, 'The truth content of the original is left behind in its history, [and] the photograph captures only the residuum that history has discharged'.[20] That is, it is only when we forget the origin of an image that it becomes an *arché*, making the archive possible (*arché* means a substance or primal element in early Greek philosophy).[21] In other words, archival images have no autonomous existence without their connection to the Zeitgeist, and we must continually reconstitute their meaning.[22] However, photographic documentation often closes the door to further interpretation. Andrew Higgott and Robin Wilson remind us of the fact that archival photographs may serve to fix an urban area and may even 'freeze' time to bring about understanding.[23] To capture a 'critical reality', the photograph must create an *arché*, while opening points of contact with each era without anchoring the image to one fixed meaning.

In the artworks on Nanjido, place signifies the accumulation of times that have been subject to the vagaries of history. Edward Casey clarifies that, amongst other philosophers, like Heidegger who forays into the history of place in his essay 'Building, Dwelling, Thinking', Foucault is the first to develop the genealogical thesis fully. For him, space and place are historical entities, subject to the flow of time.[24] In 'Of Other Spaces', Foucault specifically demonstrates that 'space itself has a history in Western experience and it is not possible to disregard the fatal intersection of time with space'.[25] It is not only true of the West, but also true of other nations that have passed though different historical experiences. Casey also points out that Foucault makes less meticulous distinctions between basic terms like 'place', 'space', 'location' and 'site'. As a result, these terms often run together, or he uses them interchangeably without precise definitions of the notions.[26] Nevertheless, it is notable that Foucault recognised and led the discussion on the idea that space is not absolute and place is not permanent but subject to the most extensive historical vicissitudes.

This concept is especially crucial to Nanjido Post-Landfill Park since the site has experienced several dramatic changes over the decades. With respect to the 'place as historical entities', however, the aforementioned photographs of Nanjido Landfill hardly push the discussion to include the site's layers of socio-economic meaning. The Landfill's essence as historical space is derived from these layered meanings, which represent the fundamental value of waste itself. The same limitations are true of the promotional photographs of Nanjido Post-Landfill Park, since they place an emphasis on the site's remediated natural environment. By doing so, they confine the viewers' understanding of the Park to a given image, rather than extending their thoughts on the site to include a broader context or providing insight into the place as an accumulation of times.

When artists conduct a photographic practice on urban development to record the present situation and evoke the past of a site, one of their major concerns is 'memory'. Ironically, however, the German aesthetic psychologist Matthew Vollgraff argues that a photograph is a kind of memory hole; it is a site of forgetting rather than one of remembering.[27] He states that our consciousness is responsible for establishing the provisional status of given configurations.[28] The photograph tends to remain conditional; subsumed into an authoritative power of stereotyped image-making, the photograph does not become an autonomous existence open to the spatio-temporal context of history.

Higgott and Wilson's critique on archival photographs contends that they often freeze time and fix the urban space, and Vollgraff's psychological analysis of photographs shares this notion. Photographs, especially those taken for archival purposes, often fail to touch upon the historical contexts that are reconstituted as memory-images in the individual conscious or unconscious. As a result, not only does the photograph remain provisional but it also becomes a frozen image of the urban situation, which can only exist in constant flux with the environmental and the social circumstance.

In this regard, we can say that most journalists' and government's archival photographs of Nanjido Landfill and Nanjido Post-Landfill Park engender a fixed image in the viewer's perceptions, resulting in the viewer's failure to reach his/her own memory-image. Therefore, rather than merely recording sheer images of scenes on a site, artists try to use different artistic devices so that the photographs can first reach the viewer's memory-image. This removes the vulnerability that results when image-making freezes a scene and ultimately demonstrates the artists' critical view on the reality.

In terms of the artistic methodology, performative practice is relatively more effective in conveying the artist's intended message than are image-based visual works, which are more likely to be misrepresented in different contexts. This is because the performative practice is often based on human-to-human communication and a direct experiential understanding of the environment and the society in which the inhabitants live—it is an active engagement with the *situation*. Accordingly, the effectiveness of performative practice is particularly true of salvaging the value of wasted lives and materials. Mierle Laderman Ukeles' decades-long landfill project in New York's Fresh Kills Landfill demonstrates an exemplary case of performative practice and its representation of an aesthetic message on the marginalised social population.[29]

So far, there has not been an artistic practice as a critical comment on and representation of the reality of the Landfill and its people during Nanjido's landfill period. In the 1980s and 1990s, it was not an aesthetic concern for artists to engage with the issues of wasted materials and the population of Nanjido Landfill. At present, it has become more difficult to explore the value of the past landfill and its inhabitants because the Nanjido Post-Landfill Park covers the Landfill and the former community has been scattered across different parts of the city. Consequently, it is harder to represent the integrity of the current Nanjido Post-Landfill Park as a place—or non-place—entailing an identity forged from its accumulated times, including the past and the present as layers of the past.

It is significant that artists who produced site-specific artworks in Nanjido Post-Landfill Park (e.g. Wonho Lee and Joon Kim, whose works of art will be analysed subsequently) began their aesthetic projects as a result of their perceptual sense of unease and the Post-Landfill Park's sense of place or placelessness. It is also notable that they used non-visual sensory apparatuses as media to represent invisible materials and the immaterial layers of history. In other words, the fact that the Landfill itself no longer visually exists has contrarily provided the artists with the potential to imagine and illustrate the sense of place or placelessness from diverse perspectives. Taking an inverted approach, they seek and unearth the root of the sense of place or placelessness through the psychological symptoms of the site's sense of unease.

Artistic exploration of Nanjido Post-Landfill Park

SeMA Nanji Residency

SeMA Nanji Residency (established in 2006) is one of the cultural facilities in Nanjido Post-Landfill Park. The residency is equipped with a main residency building and two cylindrical gallery buildings for exhibitions and performances.[30] Located on the ground level of Noeul Park (Landfill 1), it represents the Park's cultural value, one of the characteristics of a global park along with the Park's leisure facilities and programmes—they have more symbolic brand value than practical use value. The Seoul Museum of Art operates the artist residency programme. When established, it began as a programme to provide local artists with studio spaces and exhibition opportunities. Since 2012, the museum extended the programme to include overseas artists and exchange programmes with international institutions.

There is a discrepancy between the official English and Korean names: SeMA Nanji Residency in English and SeMA Nanji Art Creation Studio in Korean. The English name implies that the City planned to use the facility and its programmes for artist residencies as in other residency programmes, in which artists reside for a specific duration, short term (1 to 3 months) or long term (1 year). However, in principle, the building did not have the administrative permits for residence or overnight stays. This is why the building's Korean name does not include a word that means residence. Nevertheless, there has been a tacit agreement between the City and the museum that allows artists to reside in the building without upsetting the building regulations.

Many artist residency programmes around the world, especially those of renowned museums, aim to provide selected artists with workshop spaces and exhibition opportunities, but staying in the studio is not an absolute requirement. In the SeMA Nanji Residency's case, artists must be physically present and working in their provided studios, and foreign artists inevitably reside in their studies, too. Although many artists commute and some even work in their own studios, the museum requires all of the artists to prove their occupancy by signing an attendance book. Overall, the SeMA Nanji Residency programme is more about giving the participating artists the title of museum resident artist rather than about providing a workshop space for artists in need, but practical occupancy is ironically more necessary than in other museum residency programmes.[31]

According to the pollution expert of the Seoul Institute, it is safe enough for an adult to engage in various activities in the landfill-turned-park, but not recommendable to reside inside the Park area, especially before the Landfill's stabilisation has been completed.[32] Even though there is a formal measure of the pollution's safety levels, particularly the extent of air pollution, different people have different interpretations of the degree of safety. Some artists have expressed grievances about headaches of unidentified causes during their residency, whereas others have made no complaints of health problems and still others jog in the Park every morning during their residency. Although the Post-Landfill Park residency's environmental appropriateness is still uncertain, there have been no significant symptoms of fatal disease thus far.[33] In another sense, however, it shows that people are aware of the latent risks (the physical existence of the toxic-laden Landfill beneath the Park) cognitively and sensorially. People feel a sense of unease just knowing there is potential risk whether or not they have any health symptoms. Therefore, along with the physiological symptoms caused by the assumed pollutants (pathological dis/ease), an uneasy atmosphere surrounds the Park (sensory un/ease). Some artists in residence capture and represent this unease in their site-specific works of art.

Nanjido Post-Landfill Park's dramatic transformations through socio-historical changes have given the site unique characteristics, so the possibility for resident artists to consider site-specificity for their art projects is relatively high. However, the SeMA Nanji Residency is officially open to all sorts of subjects and genres. Since the criteria for selection is not limited to site-specific ones, only a couple of artists have engaged with site-related issues of the Nanjido region so far. Amongst the approximately 240 artists who have been in this programme since its opening in 2006 until 2016 (2016–2017), only three artists (two during the programme, one after[34]) directly deal with Nanjido Post-Landfill Park as their subject of artistic concern. To examine Nanjido Post-Landfill Park as a place or non-place, the artist's own in-situ experience of the site and insight into the diachronic axis of the history are significant. Therefore, I analyse the site-specific works of Wonho Lee and Joon Kim during their stay in Nanjido Post-Landfill Park to gain some perspective.

Site-specific art on Nanjido Post-Landfill Park: embodying the past-present

Wonho Lee and Joon Kim examined Nanjido Post-Landfill Park's sense of unease through material—as in substances, e.g. soil, air (smell and gas) and sound, as opposed to concepts—and sensory approaches. Lee particularly focuses on the olfactory sense, while Kim's major medium is auditory. They took part in residency programmes in 2010 and 2012 respectively. At that time, Nanjido Landfill was undergoing a more active process of stabilisation and regeneration than in more recent years.

Unlike relatively more renowned Western artists, who deal with the existential issues of wasted material and people living and/or working in landfills (e.g. Ukeles), Lee and Kim focus more deeply on the Nanjido Post-Landfill Park's 'place', particularly the conflicting sense of the site, which is derived from the co-existence of the interior Landfill and exterior Park. The aura is simultaneously material, socio-historical and psychological. By representing the conflicting sense of the Post-Landfill Park site, they illuminate the changing value of the material objects (waste) and the place (landfill-turned-park), making known the fact that there is no absolute waste. Moreover, they also demonstrate that the obvious yet invisible existence of the conflicting elements creates the site's sense of unease (dis/ease).

Although Lee and Kim's works take the form of video and structural installations due to the limitations in the contemporary arts' pattern of presentation, which is based on the white-cube gallery format, non-visual sensory instruments—e.g. olfactory and auditory—combined with storytelling that incorporates the artists' notes are the real foundations of their projects. In other words, Lee and Kim use 'narrative' as the communicational medium linking the non-visual senses to the spoken language in their artistic projects on the socio-historically charged place of Nanjido Post-Landfill Park. Their 'narratives' complement the figurative limits of the non-visual aspects, or the less descriptive sensory devices of these artworks. An analysis of their site-specific works will show how these artists use non-visual senses and instruments to reveal the existence of the Landfill's invisible past. It will also reveal how Nanjido Post-Landfill Park produces or removes the characteristics of place, or covers the place with placelessness.

- Wonho Lee's *(Nan)Ji*: conjuring up the past

For Wonho Lee (2010–2011), Nanjido Post-Landfill Park was a place of curiosity particularly because he had lived in the neighbouring village of Mangwon-dong during the Landfill's stabilisation period. Mangwon-dong is located 5 to 10 minutes away from the Nanjido region by public transportation. While he was living in this village, he had never actually visited the Nanjido region, but had simply acknowledged the area as the formerly largest landfill in Seoul.[35] For him, as for many Seoul citizens other than the Landfill inhabitants, Nanjido Landfill was an imaginary space that had symbolised the unsanitary, wasted, and thus, disliked site. The dichotomous idea of the sanitary-safe versus the unsanitary-threatening largely constructed or distorted the public's imaginaries of the Nanjido region during its landfill period. Lee was no exception to the preconceptions that shaped the society's perceptions of Nanjido Landfill.

Nanjido Landfill's perfect transformation into a park, the complete envelopment of the (objective) histories and (subjective) memories of the past, had surpassed Lee's imagination, and it inspired his artistic project on Nanjido Post-Landfill Park—*(Nan)Ji* (Figures 5.3, 5.4).[36]

Figure 5.3 Wonho Lee. *(Nan)Ji*. Single channel video 8 min 41 sec. Film screening exhibition view. 2011. A military training aircraft circles above the campsite on top of Noeul Park (Landfill 1). © Wonho Lee

Figure 5.4 Wonho Lee. *(Nan)Ji*. Single channel video 8 min 41 sec. Juxtaposed film stills. The artist shot the same scene of the site at different hours of the day to combine them into one film. 2011. © Wonho Lee

Above Nanjido Landfill Park, aircrafts, operated by training pilots, circle around dozens of times every day, making thundering noise. From time to time, military helicopters fly across the sky and create even more powerfully roaring sounds as if to prove that they are doing their duty. When looking up at the flight, I could easily recognise the type and code-number of the aircraft printed on the bottom of the flying vehicle. Is it because they are flying low or because Nanjido is so high? The old fetid stench of Nanjido Landfill no longer [sensorially] exists. Yes, it has been buried deep underground. Are the garbage and its smell the only things that were buried? Under the name of the Landfill's stabilisation/regeneration, *the garbage has been regenerated and turned into gas*—by way of the methane gas collecting technique—providing the neighbouring region [of Sangam-dong] with energy resources. But how can the unknown histories, buried underground and beautifully covered with the grass field, be regenerated? Disrupting the grasslands to uncover the missing histories would only end up unleashing the intolerable toxic stench of the past. The summer of 2011, Nanjido was full of excited campers and teeming with the smoke and smell of their barbeque parties all over the campsite. Deep under Nanjido, however, a *sense of unease* exists, lingering like the stale smell that saturates old clothing for a long time. *The overlapping imageries of the aircraft constantly flying above the site and the blurring [barbeque] smoke trigger a strange feeling.* Such uncomfortable feelings, residing in the corner of my mind, are perhaps fundamentally not unlike those that the [Seoul city] space outside Nanjido has long harboured within itself. Now, Nanjido attempts to provide us with an amnesic space, skilfully hiding not only the past but also the present. But I can't get away from my perpetual doubt of the stench coming out of the deep under the dark. (Italics and bracketed words added)

From the artist's notes on the project titled *(Nan)Ji* (2011)

Lee views the Nanjido region not as a site that is restricted to the single meaning or function of a designated time, either that of the Landfill or the Park, but as a site of accumulated historical times. The written description from the artist's notes, presented as a supplement to his video work, constitutes a crucial part of *(Nan)Ji*. In the comprehensive verbal-visual production, these two elements complement each other to make a single narrative statement[37] that illustrates Nanjido Post-Landfill Park's sense of place or placelessness.

Prior to his *(Nan)Ji* project, Lee had shown an interest in the basic materials or elements on site. For example, Lee used dust and charcoal powder found on the gallery floor in the work titled *Adler street projekt in situ* (2009). In *Time Exposure* (2009), he used wallpaper that had formerly existed in the space and mixed it with new wallpaper to visualise the accumulation of time; he then constructed the present by layering one period of wallpaper upon the other. For Lee, the material itself represents a facet of the spatial elements that create a place; in other words, the materials that carry the site's history constitute the site-specificity of the space and ultimately distinguish it as a place.

Nanjido Post-Landfill Park was another subject of concern in Lee's line of art projects on material and space, and he approached this work in the same way he had for years: examining and sensing the materials of the space, and extracting the contents embedded in the mixture of the material-space. In the work titled *(Nan)Ji*, Lee focused on the auditory (e.g. the sounds of the military aircraft and campers) and the olfactory senses (e.g. the smell of barbequing in the campsite). The olfactory sense particularly gave him insight into the conflicting layers of the past and present in this work, and this ultimately led him to examine the sense of unease in Nanjido Post-Landfill Park and the ambiguity between place and placelessness.

To reveal the cause of the site's sense of unease, Lee juxtaposes the conflicting phenomena of the military discipline against the Park's leisure activities, highlighting their uncomfortable co-existence. With this in mind, he shot the same scene of the site at different hours of the day to combine them into one film.[38] The aircraft, here, is a metonym that signifies the line of modernity and the contemporary history of South Korea; after the Korean War against the communist bloc, the military dictatorial regime led South Korea's economic development and modernisation. However, there is a tacit common acknowledgement that developmentalism was not in tandem with democratic advance and took place under fear. For that reason, this flying aircraft represents the symbolic fear and historical tacitness in his film. These conflated layers between knowing and unknowing (or the oppression of knowing) lead to the sense of unease. On the other hand, the people's momentary leisure activities in the Park, represented by the smell and smoke of the barbeque, the numerous camping tents, the din of voices expressing the people's enjoyment, all under the unidentified aircraft constantly flying over the Park, add another layer to the sense of unease of the Post-Landfill Park.

As for his photographic representation, Lee stood on top of the Noeul Park (Landfill 1) mound to assume the role of a subject experiencing the site, while a flying military aircraft constantly hovered above his head. His photographs with this aircraft perched above the campsite make the discomforting feelings of the place tangible. Ultimately, the unknowable—not knowing the clear identity and function of the flying object—generates the discomfort. In this way, Lee represents what Kracauer called the residuum that the history has discharged. In other words, it is not the (knowable) image but the (unknowable) image-hole of the site or *arché* that forgets its origin. Lee manages to represent a moment of contact with the space-time without anchoring the site's image to a fixed frame through his images because the historical narrative is omitted.

Another medium that leads the audience to histories and memories of the past is the smell of the campsite barbeque. What is notable is that the present savoury smell re-activates the imaginary smell of the past. Social anthropologist Uri Almagor argues that an odour of the present can be compatible with an event experienced in the past. Almagor describes the recollection of the past through odour as going backward and forward between 'time out' situations that offer a reference to the time and space of a vividly present past event. He elaborates that when recollecting odour in a certain context, the elements of space and time are indistinguishable.[39] Almagor asserts:

> The sense of smell is considered powerful, suggestive and characterised by its immediate reaction. Odours convey powerful messages that cannot be ignored […] odour is a sort of 'social enzyme' for it is a catalyst of awareness that serves to intensify our most direct experience in space and time.[40]

In Lee's case, even though he himself had never smelled the stench of Nanjido Landfill before, he acknowledges that the palatable barbeque smell of the present campsite is in stark contrast to the putrid past odour of the landfill, and that their juxtaposition lies outside the normal understanding of time-space boundaries. The present savoury smell functioned as a referent that activated the artist's recollection of the imaginary smell of the past, allowing him to go backward and forward of the two 'time out' situations where space and time are indistinguishable. In other words, the imaginaries of Nanjido Landfill and its fetid odour were so strong that the artist could employ them as instruments to draw a collective sensibility and memory of the past site. In effect, smells compress past and present times and spaces and extend beyond the linear concept of time-space.

Like the scent of the barbeque, the sound of the flying aircraft is a predominant sensory element in Lee's work. These elements give rise to the imagery and olfactory senses of the past, leading us to oscillate between the past and the present. First, the flying aircraft, an object described both as video images and written language, reminds us of images associated with South Korea's modern and contemporary history, including the military regime in power at the time of Nanjido's landfill period in the 1970s and 1980s. The military aircraft itself also represents the ironic characteristics of defence and offence; it is formally said to function for defensive purposes, yet it is simultaneously conceived as an intimidation tactic for its potential offensive power. The recurring film shots of the flying aircraft, unceasingly hovering around the site, not only blur and overlap different time zones, but also provoke fear of its potential physical power. Second, the overlapping imaginary stench of Nanjido Landfill and real scent of present-day barbeques juxtapose two geologically co-existing material presences: the Landfill and the Park. As odours are context (time-space)-bound,[41] the overlapping smells of the past and the present entail a conflation of the two different historical contexts (or time-spaces) of the Nanjido region.

In addition, the mound's noticeable height helps us to imagine the phenomenal mass of the mound, i.e. the unaltered extant garbage dump. Yet, at the same time, Lee's statement 'the garbage has been turned into gas' suggests that visible and tangible features transform into invisible and intangible traits, which may induce the confused perception that the past/landfill/unsanitariness/threat to health (pathological and psychological risk of disease) has disappeared.[42] Even though it is in the process of material transformation, the physical Landfill does in fact exist, and its historical past remains in the form of the past-present. Likewise, Lee, in his work *(Nan)Ji*, delineates how the invisibility of the past—in existence but unseen—consistently lingers as a part of the present site, thereby creating the past-present or contemporary past.[43]

• Joon Kim's *Instant Landscape*: unearthing the past

Joon Kim's (2011–2012) primary medium in his research-based art practice is sound recorded on site. In the site-specific project titled *Instant Landscape* (2012–2013), Kim integrated the sounds recorded in Nanjido Post-Landfill Park with photographs of Nanjido Landfill (closed yet unofficially in operation) that he took in 1996, 15 years before his return to the region as an artist-in-residency (Figures 5.5, 5.6). This study analyses the ways in which the artist understands the site through non-visual sensory elements (the auditory and occasionally olfactory and tactile) and synthesises these mechanical and sensory elements into his spatio-temporal and socio-historical research-based aesthetic practice. This critique aims to reveal how aesthetic practices represent the urban fabric formulated by environmental, social and human subjectivity.

Kim's artistic practice focuses on recording environmentally and socially meaningful yet unfamiliar sounds in normal audio circumstances. On the one hand, Kim collects favourable sounds of the natural environment essential to the environmental ecology. On the other hand, he captures noise, or the sound of the mechanically rendered artificial environment of modern and contemporary society, which creates a stark contrast to the sounds of nature. For example, he explores physically harmful sounds like those from electromagnetic fields (emf), while collecting the sounds of nature in historically meaningful sites, such as a water storage unit used in the Japanese colonial period.

In *Instant Landscape*, Kim collected both sounds that are presumed to be benign (the exterior Park sounds) and malign (the interior Landfill sounds). More specifically, he first gathered the micro sounds of Nanjido Post-Landfill Park, including those of insects, water and wind. He then collected the underground sounds of the Landfill in the process of stabilisation and regeneration, particularly in the area close to the methane gas collectors.

Figure 5.5 Joon Kim. Sound recording of the exterior of Nanjido Post-Landfill Park. Exterior sound (the Park) https://soundcloud.com/joon-kim-3/sound-in-nanjido. 2012–2013. © Joon Kim

Figure 5.6 Joon Kim. Sound recording of the interior (Landfill) of Nanjido Post-Landfill Park with a microphone set on the methane gas collector. Interior sound (the Landfill) https://soundcloud.com/joon-kim-3/sound-in-nanjido-worldcuppark. 2012–2013. © Joon Kim

The recordings, which included the sounds of the garbage's natural decomposition and of the artificial bio-chemical regeneration processes, form a mixture of unidentifiable hissing noises.[44]

The Landfill's interior sounds reveal a 'noise' of its own. Contrasting the sounds with those of the exterior Park's benevolent ones, and vice versa, amplifies the cacophony. Above all, this contrast reveals the existence of the invisible noise beneath the Park. This noise represents an acoustic version of the material garbage, abandoned as waste in the modern city of Seoul. Kim interprets the invisibility of the materiality of the waste, or the city's attempt to make the material waste invisible, as the 'covering of incidents' that occurred throughout Korea's modern and contemporary history:

There are diverse contrasting aspects in the process of the Nanjido region's transformation: from abandoned space to creative [cultural] space, from a forgotten space to a monumental space, from the space of a garbage dump to that of energy production, from the space of massive pollution to that of ecological reproduction. [...] By concentrating on diverse micro sounds, we can find that they imply more than different material qualities [as they reveal socio-economic and historical layers, too]. Sounds from the vibrations and wavelengths created by the bio-chemical operations inside the garbage mounds to the wind, birds, insects and electronic resonance (e.g. sounds from electric poles, the power plant and cars) enable us to imagine the *temporal duration* that links the past to the present. *Instant Landscape* aims to transform the bio-chemical reactions of the collected wavelengths, vibrations and friction originating from the gas emissions and leachate flow into sounds by means of electronic instruments. It then records and presents them [in the auditory format] to the audience. *The presentation of the recorded sounds of the Park's exterior and those of the interior—the garbage dump of Nanjido Landfill—is an attempt to interpret the ironical duality and systematic problems of South Korean society: 'the covering up of incidents' [i.e. a temporary reaction to symptoms of social issues, which neglects to resolve the cause of the problem].* Meanwhile, I coincidentally captured photographic images of the Nanjido Landfill community during the early winter of 1996 when the site existed as the Landfill, a place of social isolation. The photographic images of the past Nanjido region revive the historical imageries of the site. They are juxtaposed with the sounds of the present, extending the audiences' interest in the site to the *space of potential*. (Italics and bracketed words added)

From the artist's notes on the project titled
Instant Landscape (2011–2012)

First, by contrasting the opposing sounds from the exterior and the interior of the Landfill mounds, the artist not only emphasises the different biophysical state of the two geological layers, but also enables us to imagine the spatio-temporal differences between the two spaces. Second, regarding Kim's understanding of the 'systematic problems of South Korean society: the covering up of incidents', this work attempts to unearth what the new era's social norms have buried. Third, the artist's practice of 'noise' collection rather than a neutral sense of 'sound' collection is the foundation of *Instant Landscape*. We can view Kim's project as analogous to that of scavengers in that noise is equivalent to the waste of sound.[45]

Jonathan Sterne, while commenting on the 'audiovisual litany', states:

> Hearing is treated as the better sense since it is the 'inner' one. While seeing creates distance, focuses on the superficial, and calls on the intellect, hearing surrounds us with sounds, penetrates deep into the heart of the matter, and is inclined to the affective.[46]

Regardless of the religious overtones of this perspective,[47] the auditory mode is indeed effective in immediately conjuring up a certain temporal time-space, particularly one that is not visible or tangible.

In 1931, the physicist G.W.C. Kaye interpreted noise as 'sound out of place', which precedes Mary Douglas's definition of 'dirt' as 'any matter out of place' in 1966. He claimed that sound could lose its place by its 'excessive loudness, its composition, its persistency or frequency of occurrence (or alternatively, its intermittency), its unexpectedness, untimeliness or unfamiliarity, its redundancy, inappropriateness, or unreasonableness, its suggestion of intimidation, arrogance, malice, or thoughtlessness'.[48] Seen from this point of view, the sound of the Landfill's interior is noise for its excessive loudness, disconcerted composition, irregular frequency, unexpectedness, unfamiliarity, inappropriateness and unreasonableness. Sounds also become noise relative to the appropriate volume, concerted composition, regular frequency, expectedness and familiarity of the sounds of the exterior Park space, i.e. the sound of nature. Contrasting the interior with the sound-cleanliness-appropriateness-safety of the exterior Park further amplifies its assumed connotations of noise-dirtiness-inappropriateness-threat.

Meanwhile, *Instant Landscape* is more than just a sound recording and its presentation, as it includes a series of photographs that the artist took in 1996. Interweaving these photographs with the collected sounds makes them a part of the audio-visual historical narrative. When Kim first visited Nanjido Landfill and took these photographs in the mid-1990s, the residences of Nanjido Landfill inhabitants partially remained, and the inhabitants were protesting the City over the issues of compensation and their relocation to other regions.[49] At that time, the global economic system flowing into South Korea had suspended the Landfill's prosperity. As a result, the price of local recyclable items plummeted, ultimately causing the demise of Nanjido Landfill's economic community. This was at the same time that the Sangam-dong area was appointed as the site for the FIFA World Cup main stadium, which affected the plans for the region's redevelopment, including the Landfill closure and relocation of its inhabitants. In this historical space, Kim sensed, witnessed, experienced (consciously or subconsciously) and captured the site's state of unease and the psychology of the people's uncertainty about their future.

As discussed in Chapter 3, there are dual aspects to garbage collecting: recycling and scavenging. Recycling in the landfill is a trade for profit, or the garbage collectors' attempt to assimilate themselves to the existing social system through their economic activities. On the other hand, scavenging, in its essence, retains the characteristics of a potential threat to the existing socio-economic norm since it disrupts the so-called improper waste discarded outside the boundaries of the modern city. The scavengers' power to intimidate comes from their anomaly and their subversion of socio-economic standards. The disruptive potential of garbage collecting dwindled remarkably in Nanjido Landfill during the mid-1990s. This change imposed more layers of unease upon the site. On the surface, this sense of unease transpired from the community's anxiety over losing its economic ground; beneath the surface, it came from an undercurrent of fear that the scavengers' base for emancipatory potential was at stake, suppressed by the new socio-economic order of neoliberalism. The pile of dried corncobs, refused furniture, discarded household items and multifarious paraphernalia, presented in Kim's photographs, are indicative of the precarious status of this site (Figure 5.7). The scarecrow, set up on the Landfill slope in vain, aptly represents the destiny of waste on the

Figure 5.7 Joon Kim. *Instant Landscape*—a pile of dried corncobs abandoned in Nanjido Landfill. Photograph. 1996. © Joon Kim

verge of facing a new economic system's denial once again. Once a golden field for the garbage collectors or scavengers, the socio-economic community of Nanjido Landfill approached its termination.

Joon Kim scavenged for the Landfill's noise to bring what had been buried for its unwholesomeness back into our sensible realm, thereby exploring the co-existence of the multiple layers of materials. These layers symbolise the socio-economic contexts of different eras, which constitute the present time-space. Scavenging for noise, in this sense, not only causes sensory unease, but also evokes a feeling of threat. Since the act brings up the ominous connection between perceptual un/ease and physical dis/ease,[50] it brings to mind the zealous prohibition of contamination in the urban space under the modern ideology of sanitation.

Through their site-specific art projects, Wonho Lee and Joon Kim raised the issue of the sense of unease perceived in Nanjido Post-Landfill Park and investigated its causes by means of sensory methods, particularly the olfactory and auditory. Lee and Kim's aesthetic projects enable us to view the sense of unease as a discomforting state of mind caused by the uncertainty about the site's identity, and to discuss it within the discourse of place or placelessness. I assert that place is created by the connection to both the historical context (temporal context) and the regional context (spatial context), which Nanjido Post-Landfill Park suppresses or lacks as it has become, as Relph called, Disneyfied or museumised.

These aesthetic attempts to unearth the hidden past are perceived as threatening because they disclose uncleanliness and the fact that the society endeavoured to hide the unsanitary. The artists broke the clear-cut borderline between the sanitary and the unsanitary, and the appropriate and the inappropriate, ultimately causing disarray and disorder, as the scavengers (the garbage collectors) had in the past Nanjido Landfill site. From a spatio-temporal outlook, their artistic practices are attempts to excavate the accumulated history of the space, which is crucial to making the site a place.

Likewise, their arts of disruption inversely enable us to evaluate the ways in which the uncomfortable co-existence of the two different historical times resulted in the invisibility, intangibility and immateriality of the site's past. This reality has generated uncertainty about *the suppressed and the hidden*, which ultimately evokes the sense of unease and possibly the sense of placelessness. As mentioned before, *the suppressed and the hidden* part of the place of Nanjido Post-Landfill Park is bracketed so that it emits a sense of place[less]ness.

Artists often employ materialist methods with sensory instruments rather than with visual ones, particularly in projects that deal with the invisible waste in a closed landfill site as in Nanjido Post-Landfill Park's case. The performative art projects in operating landfills—for example, 'The Fresh Kills Landfill Project' of Mierle Laderman Ukeles—are rooted in the material contact with wasted objects or the wasted population. As opposed to the conceptual aspects, the material elements are crucial to the aesthetic study of the landfill because they inquire into the essence of the landfill itself as a mass of materials. They also question the ontological value of the landfill, as well as the physical and existential value of the landfill populations.[51]

Regarding the artistic media, Lee and Kim's choice to use non-visual sensory approaches to explore the visual absence of past materials and the presumed lack of the sense of place is notable, because it effectively reached the imaginaries of the site engrained in the people's perception. The olfactory and auditory instruments were useful in addressing the existence of the invisible or lack thereof by direct contact with these perceptions. The two artists' projects are also a psychological examination of the site because they deal with the invisibility of the past time-space, the social authority's tendency to suppress it and the ways in which the historical (temporal) disconnection between the past and the present creates the Park's sense of unease and place[less]ness. Their projects focused more on the historical (temporal) relations than on the regional (spatial) relations, both of which, I argue, are preconditions for the making of place.

The material and sensory approaches are interconnected by their mutual concern with the restoration of the sense of place: first, through their rediscovery of material value; and second, through their recognition of the absence or suppression of the sense of place. Here, the process of salvaging materials and their values recovers the historical layers embedded in them, too. Drawing on Foucault's ideas in 'On Other Spaces', these actions embody the process of retrieving a place—the heterotopia of reality—from the idealised yet placeless utopia of Nanjido Post-Landfill Park. That is, these kinds of artistic projects produce epiphanies of a place of historicity, which, after the Park period, has become a mirrored image.

We can view the artists' engagement with the matter of waste and the landfill in the broader context of the environmental and social ecologies. Essentially, it begins as an environmental concern, particularly for the deteriorated conditions of nature, and grows into a socio-ecological concern, as the waste treatment in the landfill is inseparable from the lives and works of the Landfill's inhabitants and neighbours. While some artists illuminate environmental issues as a planetary concern, others focus more closely on the characteristics of the landfill site and the people, including the cases of closed landfills as in Nanjido Post-Landfill Park. The works of both groups similarly disrupt modern society's norms of industrialisation, developmentalist urbanisation and the orderly categorisation of affairs. In other words, they attempt to integrate the abandoned and marginalised aspects of society into what Mierle Laderman Ukeles calls the 'urban flow' that links the earth (the natural environment) and the city (the built environment). These attempts to re-explore the objects and people positioned in the pre-organised set of urban design are, by the modern *norm*, acts of menace that may cause confusion and disorder. Here is the disruptive power of aesthetic practices; other disciplines based on the logic of modern science can hardly produce the same effect.

Notes

1 The socio-economic position of the Nanjido region's surplus population during and after the landfill period can be explained through Zygmunt Bauman's
   research on multiple aspects of wasted lives, particularly in the global era.
   Bauman extensively deals with the analogy between material and human waste
   in the globalised socio-economic context. For him, the existing society regards
   the significantly increasing number of immigrants and refugees (a proportion
   of asylum seekers created by either war or natural disasters) as surplus. See
   Bauman, 2012 [2004]: chapter 3.
2 Art History categorises modern art in the early twentieth century as historical
   avant-garde, distinguished from progressive art movements in later years, particularly the conceptual art after the 1960s. These artists attempted to break the
   art conventions of the previous era that had become co-opted by the art market
   in the mid-twentieth century. The conceptual art that overlapped with minimal-
   ism in the 1960s and 1970s empowered the artists to practise socio-politically
   engaging art that deals with subjects that include gender and ecology. We may
   be able to call them post-conceptual artists.
3 Benjamin Buchloh describes art projects derived from bankruptcy, particularly works dealing with commodified objects, as 'a new "phenomenology"
   of advanced reification' or 'the trashing of trash' that mimetically follows 'the
   universal condition of the commodity' in the globalised era, in which market
   structures depend upon the ever more rapid pace of compulsory obsolescence
   in which the 'new' is produced as already-obsolete detritus (Benjamin Buchloh,
   'Detritus and Decrepitude: The Sculpture of Thomas Hirschhorn', *Oxford Art
   Journal*, Vol. 24, No. 2, 2001: 48, 55). Artists, working with objects considered
   obsolete, take a position that not only stridently criticises consumer capitalism,
   but also simultaneously challenges the idea that the consumed commodity exists
   as permanent detritus, or the always-already obsolete. Instead, these artists claim
   the universal condition of objects is cyclical in nature. The cyclical interrelationship of construction-demolition in the spatial dynamism generates the logic of
   production-consumption. In short, there is no such thing as absolute surplus; the
   value of commodities lies only in the relative conditions of the socio-economic
   context. See Abraham Cruzvillegas' sculptures, which he created by improvising
   with temporarily abandoned materials exhibited at Turbine Hall, Tate Modern
   (2015).
4 Mierle Laderman Ukeles carried out performative art projects, including *Touch
   Sanitation* (1978–1984) and *Flow City* (1983–1986), as a landfill resident artist
   in the Fresh Kills Landfill in New York. See Mierle Laderman Ukeles, 'A Journey:
   Earth/City/Flow', *Art Journal*, Vol. 51, No. 2 on Art and Ecology, Summer,
   1992: 12–14 and *Mierle Laderman Ukeles: Seven Work Ballets*, Kari Conte ed.,
   Bristol: Arnolfini, 2015. The 'performative', here, means the performance-based
   practice, through which the artist engages with the dynamism of social lives.
5 Drawing on Lacanian psychoanalytic and neo-Marxian perspectives mainly
   addressed by Cornelius Castoriadis, Ben Campkin argues 'adapted to an urban
   context the notion of "place imaginaries" can usefully articulate the ways that
   contested sites are constructed, recognised or distorted, from multiple and con-
   flicted perspectives, through forms of representation that are not passive but
   have agency and are affective within urban change as they engage with particular
   empirical features and material conditions' (Ben Campkin, 2013: 9–10; Lacan,
   2003 [1966] and Castoriadis, 1987 [1975]).
6 Claire Bishop, *Artificial Hells: Participatory Art and the Politics of Spectatorship*,
   New York: Verso, 2012: 278. On 'transversality', see Guattari, 2000 [1989].
7 Bishop, 2012: 284.

8 Montesquieu, in his book of political theory titled *The Spirit of the Laws*, uses the term 'uneasiness' as opposed to 'tranquility' when writing of a state or of the individual citizens. When discussing despotisms such as that in China, he mentions that they pursue 'public tranquility' as their 'aim', if not their 'object'. Meanwhile, in his discussion on the government of England, he states that it pursues the individual citizen's 'tranquility of mind'. It is not surprising that, unprompted by genuine peril or even false alarm, he should nonetheless 'fear [crainte] the escape of a good' that he 'feels', that he 'hardly knows', and that 'can be hidden from us', and that this 'fear [crainte]' should 'always magnify objects' and render him 'un-easy [inquiet] in his situation' and inclined to 'believe' that he is 'in danger even in those moments when' he is 'most secure' (Montesquieu, 1757, 3.19.27: 575–576; Paul Rahe, *Montesquieu and the Logic of Liberty: War, Religion, Commerce, Climate, Terrain, Technology, Uneasiness of Mind, the Spirit of Political Vigilance, and the Foundations of the Modern Republic*, New Haven: Yale University Press, 2009: 99–102; Montesquieu, *De l'Esprit des* lois, 1757).

9 Yi-Fu Tuan, *Space and Place: The Perspective of Experience: The Perspective of Experience*, MN: University of Minnesota Press, 1977: 6.

10 Liz Taylor, 'Place: An Exploration', *Teaching Geography*, Vol. 30, No. 1, Spring 2005: 14.

11 Edward Relph, *Place and Placelessness*, London: Pion Books, 1976: 48.

12 Doreen Massey, 'A Global Sense of Place' reprinted in *Reading Human Geography: The Poetics and Politics of Inquiry*, T. Barnes and D. Gregory eds., London: Arnold, 1997: 322–323.

13 Ibid.,: 323.

14 Marc Augé, *Non-Places: An Introduction to Supermodernity*, London: Verso, 2008 [1995]: vii–xxii.

15 Ibid.: xvii, xxii.

16 Ibid.: 96.

17 Quynh Vantu, a 2016 SeMA Nanji Residency international artist, pointed out the site's separation from other urban areas despite its location close to the bustling university towns and downtown area. She described her impressions of the Park, including the artist residency area, as a 'place of placelessness', regarding that quality as a very strong character of the site. From the interview with Quynh Vantu (7 March 2017).

18 Edward Relph identified landscapes made for tourists and entertainment districts as representative cases of placelessness. He grouped them with pseudo-places like 'Disneyfied' and 'museumised' places (Relph, 1976: 118–119). In his 'manifestation of placelessness', Relph grouped the characteristics into four categories: 1) other-directedness in places; 2) uniformity and standardisation in places; 3) formlessness and lack of human scale and order in places; 4) place destruction; and 5) impermanence and instability of places. The symbolic value and function of Nanjido Landfill Park as a monument for environmental regeneration and its relationship with placelessness is undeniable.

19 South Korean photographer Dong-hoon Sung took photographs of the landfills in Cebu, Philippines (called Inayawan, which means 'the deported or the ostracised', and closed in December, 2016) and Bangtargebang Landfill in Indonesia. Kim held a solo exhibition of these photographs titled 'A Secret Paradise' (Gallery ryugaheon, Seoul, 2015, www.photomarketing.co.kr, uploaded on 19 November 2015, accessed on 1 May 2018). Chinese photojournalist Wang Jiuliang took pictures of numerous landfills around Beijing in 2008–2009 (Eric Hilaire, 'Photographer Zooms in Beijing's Waste', www.theguardian.com, uploaded on 26 March 2010, accessed on 1 May 2018). Wang also directed a film titled *Plastic China* in 2016. In his 2015 photographs, Spanish documentary

photographer and filmmaker David Rengel captured scenes of the landfill and child labour in the Anlong Pi Landfill in Cambodia, which has become a tourist site where visitors flock to take photos of the lives of 'others' (Sarah Gordon, 'Shocking Images Show Horror of Child Labor at Toxic Cambodia Rubbish Site', www.dailymail.co.uk, uploaded on16 March 2015, accessed on 1 May 2018).

20 Siegfried Kracauer, *The Mass Ornament: Weimar Essays*, trans. by Thomas Levin, Cambridge, MA: Harvard University Press, 1995: 55.

21 'arché', *Merriam-Webster dictionary*.

22 Matthew Vollgraff, 'The Archive and the Labyrinth: On the Contemporary Bilderatlas', *October* 149, Summer 2014: 146. Ernst Gombrich states that 'stars [...] are neither beneficent nor maleficent in themselves but derive their meaning from the context' (Ernst Gombrich, *Aby Warburg: An Intellectual Biography*, London: Phaidon Press, 1986: 199).

23 Timothy Wray and Andrew Higgott eds., *Camera Constructs: Photography, Architecture and Modern City*, London: Ashgate, 2012: 18. See also Robin Wilson, 'At the Limits of Genre: Architectural Photography and Utopic Criticism', *Journal of Architecture*, Vol. 10, No. 3, 2005: 265–273.

24 Edward Casey, *The Fate of Place: A Philosophical History*, CA and London: University of California Press, 1998: 297–298; Michel Foucault, 'Questions on Geography', an interview that appeared in the Marxist geographic review *Hérodote* in 1976 and was reprinted in *Power/Knowledge: Selected Interviews and Other Writings 1972–1977*, C. Gordon ed., New York: Pantheon, 1980: 69.

25 Michel Foucault, 'Of Other Spaces', trans. by J. Miskowiec, *Diacritics*, Spring 1986: 22.

26 In many cases, the terms may not be compatible, but Foucault has set space, place and site side by side in one sentence; 'heterotopia is capable of juxtaposing in a single place several spaces, several sites that are in themselves incompatible' (Casey, 1998: 25).

27 Vollgraff, 2014: 143.

28 Kracauer, 1995: 62.

29 As a landfill resident artist in New York's Fresh Kills Landfill, Mierle Laderman Ukeles led the performance art 'Touch Sanitation' (1979–1980), notable for Ukeles' use of the first or second person perspective by engaging with the working environment of the landfill workers and not viewing it from the third person humanitarian point of view. Through the long-term performance of shaking hands with individual landfill workers, she explored their value and that of their profession, which are regarded as inappropriate for the sanitary norm of society. Her emphasis on the landfill as a site of value to be re-created as a part of the urban flow is rooted in the idea that the cyclical flow between the earth and the city embodies their relationship with one another (Ukeles, 1992: 12–14). Lucy Walker and Karen Harley's film *Waste Land* (2010) explores the value of human beings and their lives in the landfill. The film focuses on a renowned artist from the Jardim Gramacho Landfill in Brazil, Vic Muniz, who created portraits of the landfill people by using the landfill's found materials. By illuminating Muniz's growing reputation and wealth, a result of his entrance into the mainstream art market in New York, the film dramatises the myth of a genius escaping the ghetto and ascending the social ladder of capitalism. However, it also conveys the message that the landfill as a wasteland is not the termination of material or human value but a mass of potential. The creation of the landfill residents' portraits also contributes to re-explorations of the value of individual human beings in the landfill.

30 These structures were originally built as leachate treatment facilities. After a more technically advanced facility was installed, these buildings were left unused

for years. In 2006, Seoul City (Mayor Lee Myung-bak, 2002–2006, the tenth elected President of South Korea, 2008–2013) and the Seoul Museum of Art (Ha Chong-hyun, director of SeMA, 2001–2006) co-initiated the establishment of the SeMA Nanji Residency. Such environmental or cultural policies generated symbolic value, contributing to the increased positive images of the politicians and their career.

31 Interview with Jade Keunhye Lim (12 April 2014).
32 Interview with Dr Woon-soo Kim (22 April 2014).
33 Interviews with Hyun-A Cho (17 April 2014) and Joon Kim (3 August 2014).
34 In 2015, after 7 years in the SeMA Nanji Residency Programme (2008), Hyun Na produced *Babel Tower Project—Nanjido*, which was shortlisted on the Artist of the Year Korea award. In this work, the artist overlapped the Tower of Babel with a gigantic mound of Nanjido Landfill under the theme of infinite human desire; while the former is about the human desire to challenge God, the latter represents the human desire for industrialisation—constant production and consumption. Hyun Na's project, in particular, attempted to link the discourse of the environmental ecology with that of the social ecology ('An Interview with Na Hyun on the Work "Babel Tower Project—Nanjido": Shortlisted Piece of the Artist of the Year, South Korea', *Kookmin Ilbo* [16 August 2015]).
35 Interview with Wonho Lee (2 March 2017).
36 The SeMA Nanji Residency does not require the resident artists to do site-specific projects. Wonho Lee's work and Joon Kim's project are rare cases. The artist himself, however, does not see his work as particularly site-specific; rather, he views it as a production of his daily life. It eventually became the first of his site-specific pieces created in South Korea. From then on, the notions of space and place have become crucial subjects in his works. 'An artist is often influenced by his/her own works. Sometimes, the artist gets strong feedback through the working processes, which often pushes them in another direction'. From the interview with Wonho Lee (2 March 2017).
37 According to the film and literary critic Seymour Chatman, narrative is comprised of a content plane (called 'story', containing information and experience created by the author) and an expression plane (called 'discourse', containing the author's ideology and political/cultural position). The expression plane has a set of 'narrative statements' that the author intends to convey to the viewer. Chatman elaborates that the narrative statement comes in different forms depending on the genre of art (e.g. a series of film shots, a whole paragraph of a novel, or a single word) (Seymour Chatman, 'Voice', *Narrative/Theory*, David Richter ed., New York: Longman, 1996: 161).
38 Interview with Wonho Lee (2 March 2017).
39 Uri Almagor, 'Odors and Private Language: Observations on the Phenomenology of Scent', *Human Studies*, Vol. 13, No. 3, 1990a: 266.
40 Ibid.: 264–266. Almagor describes the traits of odours and the ways in which we sense them: 'The odorous mode of time and space is usually dormant, for most odours that reach us are compatible and mingle with the environment from which they emanate, and thus, perhaps, we do not notice them. But we are stimulated by an odour alien to its immediate environment'. People sense the pernicious odour of the landfill and the flavoursome smell of the barbeque as more than salient for their mutually alienated characteristics and relation to their immediate or neighbouring environment. It is more so because the Nanjido region is located close to the heart of Seoul, not on the outskirts of the urban area, where most of today's landfills and campsites are situated.
41 Uri Almagor, 'Some Thoughts on Common Scents', *Journal for the Theory of Social Behaviour*, Vol. 20, No. 3, 1990b: 189.

42 This recalls Bauman's idea of 'liquid modern' influenced by Marx's phrase 'all that is solid melts into air'. Bauman began to use the term 'liquid' as opposed to 'solid' to represent mobility and change. He expressed the transition from modernity to postmodernity as one from solid modernity to liquid modernity. See Zygmunt Bauman, *Liquid Modernity*, Cambridge: Polity Press, 2000.

43 The concept of 'contemporary past' emerged in the discipline of archaeology and anthropology to investigate the distant past onto the present. See Victor Buchli, *Archaeologies of the Contemporary Past*, London and New York: Routledge, 2001.

44 He also recorded the emf sound of Nanjido Landfill Park's exterior area, which implies the ambient yet invisible hazard surrounding the Park environment.

45 In Egypt, scavengers are known as zabaline predominantly consisting of Coptic Christians. In Mexico, scavengers are called pepenadores, and they are unionised and even powerful in their social space (Rathje and Murphy, 1992: 40).

46 Bijsterveld and Rodaway attempt to overcome this hierarchy of senses with religious overtone. Rodaway does not view the new auditory mode as 'a revival of something long since lost, but as yet another redefinition of the role of the sense of hearing […] in geographical and social experience'. See Karin Bijsterveld, *Mechanical Sound: Technology, Culture, and Public Problems of Noise in the Twentieth Century*, Cambridge, MA and London: MIT Press, 2008: 12; Paul Rodaway, *Sensuous Geographies: Body, Sense and Place*, London and New York: Routledge, 1994: 114; and Jonathan Sterne, *The Audible Past: Cultural Origins of Sound Reproduction*, NC: Duke University Press, 2003.

47 Sterne explains that while the eye (the dead letter) fills the role of the fallen angel, the ear (the living spirit) embodies our future paradise (Bijsterveld, 2008: 12).

48 Bijsterveld, 2008: 240; G.W.C. Kaye, 'The Measurement of Noise', *Proceedings of the Royal Institute of Great Britain* 26, 1931: 435–488.

49 Interview with Joon Kim (3 August 2014).

50 In discussing the socio-ecological position of Rachel Carson in the masculine field of science, Lorraine Code uses the relationship between these two. She asserts that we need to understand something of the *dis*-ease—the *un*-ease of that position—with its sexed/gendered/sexuate sources despite these categorisations lack of visibility in her time. In her milieu, sexuate thinking had not yet travelled as in ours (Lorraine Code, 'Manufactured Uncertainty: Epistemologies of Mastery and the Ecological Imaginary', *Relational Architectural Ecologies: Architecture, nature and subjectivity*, Peg Rawes ed., London: Routledge, 2013: 83–84).

51 Hilary Powell examines and demonstrates the issues of the urban space through her art projects. She focuses on the simultaneous characteristic of building and demolishing by conducting an in-depth investigation of basic materials (e.g. zinc, brick, aluminium, asbestos, steel, concrete, wood, slate and copper). By delving into the world of materials—the bio-chemical characteristics of materials and their uses in certain context—she provides the ontological base of the materials and re-vitalises the value of individual materials that were once abandoned. See Hilary Powell, *Urban Alchemy* (2015).

Conclusion

The fields of architecture and urban studies have rarely dealt with the issue of the landfill as a primary subject of research on its own, but have mentioned it as an issue related to sustainable design, which has become a planetary concern. Peter Droege states that the silence on the issue of oil dependence and its environmental consequences has built the modern histories and theories of Western architecture.[1] Responding to the environmental problems of modernisation, as Ana Miljački argues, architectural criticism has demonised any entanglement of architecture with non-recyclable consumption since the high industrial era.[2]

Under such circumstances, this study makes sense of the landfill and landfill-turned-park from an urban ecological stance that suggests a revisioned understanding of the habitat both for human and non-human entities. It also calls the demarcation between natural and artificial environments into question. Since our man-made landfills permeate the earth, they become a single entity through bio-chemical and bio-physical processes. Consequently, they materially and symbolically represent how the natural environment absorbs structures made from human intentions. In light of Richard Ingersoll's idea that buildings and their natural environment are both part of an interdependent and historically contingent ecology, we can also see the landfill as a structure dependent on the network of this operation.[3]

This research examines Seoul's Nanjido Landfill (1978–1992) and its transformation into Nanjido Post-Landfill Park (The World Cup Park, 2002–present). It describes Nanjido Landfill from the point of view of urban sanitation, defining it not only as a waste dumping site but also as a human habitat. Then, it explains the landfill's transformation into a park as a variation of sanitation from the perspective of environmentalism. In this regard, revitalising the area was one of the environmental and social phenomena that emerged in the globalised neoliberal economic system. As a study of the urban ecology, this research is chiefly grounded on the discourse on the relational ecologies of the environmental, social and human subjectivity. The study is further premised on the capitalist urbanism, from the industrial and post-industrial era during the late twentieth century to the globalised neoliberal economic circumstances at the turn of the twenty-first century onwards.

As the first English account of Nanjido Landfill and its transformation into Nanjido Post-Landfill Park written from the urban ecological perspective, primary sources played a large role in developing the research for this study. Textual and visual materials along with interviews obtained through empirical research during the fieldwork in Seoul were major components. I draw key ideas regarding the urban ecology of the site from these sources, and my analysis of them plays a crucial role in establishing the arguments of this study. The historical chronology of Nanjido, pieced together from these primary sources, mainly frames the structure of the study, and the two sections chronicling the Nanjido Landfill period and Nanjido Post-Landfill Park period largely comprise the bulk of this research.

Chapter 1 reviews the historical dynamics of the political, socio-economic situations and their impact on the changing environment of the Nanjido area. The chronological examination of Nanjido's transformations demonstrates how Seoul City managed material waste or socially inappropriate entities to meet the sanitary norms of the modern city during different political and socio-economic eras. In summary, it functioned as a site for disliked facilities, including the war orphanage, and as a municipal solid waste landfill. Accordingly, these roles stigmatised and isolated the region for decades.

Chapter 2 investigates how Seoul City or South Korea's national/municipal authority systematically categorised all things and human beings into the inappropriate (waste) and the appropriate (non-waste). Zygmunt Bauman's discussion on the 'waste management' in the making of a modern city and his extended discourse on waste in the era of globalisation form the foundation of this chapter. In the post-war space of Seoul during its high industrial era, control over the inappropriate shifted from institutional, physical control to control through non-control under the urban planning policies of the new economic system. I also examine the city's management of the urban space's cleanliness through DDT fumigation since it was one of post-war Seoul's most significant methods of city-wide sanitary maintenance. Here, I emphasise how the national/municipal authority sought to control any source of pollution or disease in the modern urban space. This examination demonstrates several important factors. First, when the concept of sanitation functions as an ideology, the social authorities, either political or corporate, galvanise sanitation policies to realise their socio-political goals (e.g. the anti-communist ideology of South Korea in the 1960s and 1970s). Second, the wide use of DDT was a global phenomenon particularly related to warfare (e.g. WWII and the Korean War), and the United States' policies on its use or prohibition had a decisive impact on those of other nations. Changes in DDT policies also increased environmental awareness on a global level. Due to South Korea's close political relation with the United States since the Cold War era, the United States' policies directly affected South Korea in diverse aspects, including its use and ban of DDT. Even though Rachel Carson's claim for environmental awareness had heavily influenced the ban on toxic chemicals in the United States, South Korea accepted the government policy without taking Carson's deeper message on the ecological feedback loop into consideration. These local and international sanitary controls demonstrate that when sanitation becomes a dominant criterion of a society, it challenges the public's understanding of the ecology.

Chapter 3 examines Nanjido's inhabited landfill where over 4,000 people resided and worked as garbage collectors, a unique case in South Korea's history. While studying landfills in the context of the relational urban ecology, we must look into the environmental and social conditions of the inhabited landfill. This sheds light on the relationship between the material waste and immaterial waste (e.g. human beings of marginalised social groups), and help us to view the inhabited landfill as a metonymy for the dual layers of waste. Specifically, the study on Nanjido Landfill as a habitat investigates the precarious living conditions within the landfill area, and the residents' discriminated position within Sangam-dong town and Seoul City. These examinations show how the society identifies the landfill inhabitants with the material waste and segregates them and the zone of their residence from the other parts of the city. Their segregation is both spatial and psychological, or material and immaterial. However, the Nanjido Landfill's location on the border endows it with duality since it positions the inhabitants' residences between the legal and the illegal and situates the residents' work of garbage collecting between productive recycling and disarraying scavenging. These border characteristics give us a better grasp of the disruptive potential within the abandoned environmental and social entities of the inhabited landfill. In essence, this underlying capacity for disorder challenges the existing social norms and disequilibrium of the urban ecology.[4] In this discussion, I use the stench from the landfill as a sensory medium to represent the dual nature of the landfill. On the one hand, controlling the odour represents the secured sanitation of the [non-wasted] urban space; on the other hand, the odour itself threatens the norm of the society by crossing the border between the waste and the non-waste.

Dedicated to the transformation of Nanjido Landfill into Nanjido Post-Landfill Park, Chapter 4 analyses detoxification and aestheticisation, key aspects of the landfill regeneration construction, revealing that regeneration relies on making the material waste and historical times invisible. Especially in relation to the international sporting event of the 2002 FIFA World Cup Korea/Japan and to the overall Sangam-dong regional redevelopment plan, I conclude that the Landfill's transformation into the Post-Landfill Park is a new form of waste treatment embedded in the urban planning of the neoliberal economic system. Transforming the unused sites of industrial eras into environmentally responsible ones equipped with leisure facilities (e.g. facilities for cultural and sporting activities) is a global phenomenon. I categorise Nanjido Post-Landfill Park as a park in the global style because it was built within a neoliberal economic system in which the Park's natural environment, cultural activities and so forth turn into symbolic exchange value, which mainly serves economically privileged groups and further marginalises the lower income class. I generally refer to the benefactors of the Park's new facilities as 'the public', which I identify as the upper middle class in South Korea's context. Also, within my interpretation of the global style of parks, I have discussed global environmentalism as a counterattack against the deterioration of nature, a result of over-production and over-consumption, particularly of the late twentieth century. In South Korea, attention to the environment increased in close connection to the nation's socio-economic globalisation in the 1990s. Since then, terms like 'environmentalism', 'environmental-friendliness', 'ecology', and 'sustainability'[5] have been used interchangeably, commonly indicating the human-centred sustainable development that lacks a relational understanding of environmental and social ecologies. This also calls the environmental activism that prioritises the preservation of nature into question,[6] since such efforts merely turn into exchange value and become co-opted by the neoliberal economic system.

Tracing the history of the site and unearthing the ecological conditions buried beneath the regenerated landscape are crucial aspects of this study, so the historical approach is the primary research method. I based the chronology's importance on the hypothesis that the sense of unease perceived in the regenerated Nanjido Post-Landfill Park, which is presumed to cause the sense of placelessness, is rooted in the layered yet covered, and thus, invisible history of the site. Chapter 5 looks into the relationship between the sense of unease and the place in an attempt to identify Nanjido Post-Landfill Park as a landfill-turned-park with the closed landfill mounds remaining under the park. For this, I use the aesthetic approach, a critique of site-specific works of art, particularly those based on non-visual senses, such as the olfactory and the auditory. Through their art practices in the Park, the artists reveal the ambiguous sense of potential threat from the extant but veiled, and thus, invisible garbage, i.e. the dirt that the society conflates with disease. Here, the idea of dis/ease is equivalent to un/ease, and the sense of unease may lead to the sense of placelessness. As these art projects represent buried material entities and annihilated histories, I assert that they are aesthetic attempts of disruption. Moreover, as Nanjido Post-Landfill Park constantly oscillates between a sense of place (the practical activities in the Park) and that of placelessness (the spatio-temporal disconnectedness of the Park), I conclude that it is a site of the sense of place[less]ness.

Meanwhile, throughout this research, I mention the toxic odour as a symbolic object that the national/municipal authority has endeavoured to control in the name of sanitation or in the effort to meet the socio-economically appropriate conditions of the modern urban space. Although invisible and airborne, the odour creates a border between the zone of the inappropriate (landfill space) and that of the appropriate (non-landfill space) in the city. Ironically, however, since it is invisible and airborne, the odour freely crosses the border that it has created, threatening the psychological security determined by the urban space's standards for cleanliness. We can perceive the threatening potential of the odour as analogous to the scavenging aspect of garbage collecting because it, too, has disruptive power over existing social norms. What's more, the artists mentioned in Chapter 5 appropriate the same method of disruption to excavate the site's buried history and revive the place where the scavengers' lives unfolded. Essentially, the simultaneous bordering and un-bordering quality of odour signifies its rejuvenating and/or rebellious power. Whether it could be self-rejuvenating potential (i.e. recovering the subjectivity of the individuals) or a cause of confusion depends on the society's understanding of the relational aspect of the urban ecology.

To summarise, approaching the subject of Nanjido Landfill and landfill-turned-park of Nanjido from the urban ecological perspective redirects the focus of landfill research from the quantitative studies of pollution levels and engineering methods of landfill management or its regeneration, to the relational urban ecological concerns of architectural and urban history and theory. Especially in the local context of South Korea where ecology is understood as the preservation of nature for anthropocentric sustainable development, the expanded interpretation of ecology could provide the foundation for a fundamentally revised perspective on architectural and urban projects.

Additionally, as the first study on the urban ecology of the inhabited landfills of the Northeast Asian region, particularly of South Korea, this research extends the regional boundaries of ecological studies. Thus far, this academic field has only focused on cases in the advanced Anglo-American regions and in underdeveloped countries where inhabited unsanitary landfills are still in operation. South Korea, a relatively economically advanced nation that is behind in terms of its socio-cultural development, could be the parameter through which we could evaluate how ecological equilibrium is accomplished or destroyed on a global level.

Lastly, research grounded on empirical knowledge, especially in the fields of the urban and the ecology, would enable academia to engage more actively with the practical issues at hand, e.g. the planetary issue of surplus,[7] or the overflowing material waste and wasted populations, including the unemployed,[8] and illegal immigrants and refugees, produced either by political and/or environmental reasons. As the most recent example affecting environmental issues on a planetary level, China's decision to cease imports of hazardous waste can be taken into consideration. In July 2017, China notified the World Trade Organization that it intended to ban imports of trash, mainly from the United States, Canada, European countries and Australia.[9] Since China had processed at least half of the world's exports of waste paper, metals and used plastic (7.3 million tonnes in 2016, according to recent industry data), its ban on trash imports has made most Western countries and economically prosperous developing countries, including South Korea, look for other outlets for their trash. As a result, these nations have sought cooperation from countries in South and Southeast Asia (e.g.

India, Vietnam, Malaysia and Indonesia; South Korea is one of the candidate import countries, too[10]). They have also made entreaties of countries in Africa and Europe (e.g. Turkey and Bulgaria).[11] This shows that waste management has become an issue that influences every corner of the globe and that the waste is likely to go to countries of the Global South and the surrounding oceans of that region, which is, as Bauman comments, an extension of the search for local solutions to globally produced problems. To stop this vicious cycle and achieve ecological equilibrium, we must increase awareness of ecology as a way of inhabiting well and guarantee the literal equal distribution of the right to this ecology on a global level. In this dire situation, the disciplines of architecture and urban studies must acknowledge their environmental responsibility, not for sustainable development through technological advancements, but for the secure and smooth operation of the environmental and socio-ecological feedback loop.

Notes

1　C. Greig Crysler et al. eds, *The SAGE Handbook of Architectural History and Theory*, London: SAGE Publications Ltd., 2012: 28, 590–601.
2　About Ana Miljački's argument, see Crysler et al. eds: 184–197.
3　Crysler et al. eds: 29, 501–512, 573–589. In a similar context, Antoine Picon states that we now use technology as a field of 'quasi-objects' dependent on networks, which is a landscape punctuated by transitory points of access and interface.
4　Squatting in and around the landfill and getting free food, clothes and housing materials from the garbage dump are regular occurrences in inhabited landfills like Nanjido Landfill; these activities challenge the ongoing consumerist economic system that produces the new as already-obsolete detritus (Benjamin Buchloh, 'Detritus and Decrepitude', *Oxford Art Journal*, Vol. 24, No. 2, 2001: 48, 55).
5　Corporations especially employ the term 'sustainability'; however, while they retool their corporate identities as socially responsible corporations, their role in environmental devastation actually increases (C. Greig Crysler et al. eds.: 28).
6　Félix Guattari suggests 'ecology must stop being associated with the image or a small nature-loving minority or with qualified specialists' (Félix Guattari, *The Three Ecologies*, London and New York: Continuum International Publishing, 2008 [1989]: 35).
7　See David Harvey's explanation of surplus in capitalism (David Harvey, *Rebel Cities*, New York: Verso, 2013 [2012]: 5, 136–138) and Zygmunt Bauman's interpretation of surplus or what he calls 'the wasted' in the globalised neoliberal economic system (Zygmunt Bauman, *Wasted Lives*, MA: Polity Press, 2012 [2004]: chapter 3).
8　As in other industries, as the presence of human labour became a nuisance in the waste management industry, operations at dumps and landfills grew increasingly mechanised (William Rathje, Cullen Murphy, *Rubbish! The Archaeology of Garbage*, New York: Harper Collins, 1992: 43).
9　The official announcement on the plastic ban came into effect on 1 January 2018. China's ban covers imports of 24 kinds of solid waste, including unsorted paper and the low-grade polyethylene terephthalate used in plastic bottles, as part of the country's broad clean-up effort.

10 China's trash ban also heavily affected South Korea in several different aspects. The national/municipal governments have sought alternative waste management methods as western countries do. Yet, at the same time, South Korea is one of the candidate import countries that may accept recyclable waste from western nations. In addition, the country has to deal with pollutants hailing from newly built incinerators on the east coast of the Chinese continent in the near future.

11 Kimiko de Freytas-Tamura, 'Plastics Pile Up as China Refuses to Take the West's Recycling', *The New York Times* (11 January 2018); Matthew Taylor, 'Rubbish Already Building up at UK Recycling Plants due to China Import Ban', *The Guardian* (2 January 2018); 'China Slows Garbage Imports and Impact is Felt Globally', *The New York Times* (25 November 2017); 'China's Trash Ban Forces Europe to Confront Its Waste Problem' (from 'Getting Wasted' series), *Politico EU* (21 February 2018).

Bibliography

Books & journal articles

Abraham, Robert D. 'Spinoza's Concept of Common Notions'. *Revue Internationale de Philosophie*, Vol. 31, No. 119/120 (1/2), SPINOZA (1977): 27–38.

Agamben, Giorgio. *Homo Sacer: Sovereign Power and Bare Life*. CA: Stanford University Press, 1998.

———. *State of Exception*. IL: University of Chicago Press, 2005.

Alker, Sandra. Joy, Victoria. Roberts, Peter. Smith, Nathan. 'The Definition of Brown Field'. *Journal of Environmental Planning and Management*, Vol. 43 (2000): 49–69.

Almagor, Uri. 'Odors and Private Language: Observations on the Phenomenology of Scent'. *Human Studies*, Vol. 13, No. 3 (1990a): 253–274.

———. 'Some Thoughts on Common Scents'. *Journal for the Theory of Social Behaviour*, Vol. 20, No. 3 (1990b): 181–195.

Apter, Emily. 'The Aesthetics of Critical Habitats'. *October*, Vol. 99 (Winter 2002): 21–44.

Ascher, Kate. O'Connell, Frank. 'From Garbage to Energy at Fresh Kills'. *The New York Times* (15 September 2013).

Assaad, Ragui. 'Formalizing the Informal? The Transformation of Cairo's Refuse Collection System'. *Journal of Planning Education & Research*, Vol. 16, No. 2 (1996): 115–126.

Augé, Marc. *Non-Places: An Introduction to Supermodernity*. London: Verso, 2008 [1995].

Bae, Sang-pil. 'A Study on the Policy Making Process Regarding the Environmental Recovery of Nanjido'. Master's thesis in Urban Environment Policies at University of Seoul (Graduate School of Urban Sciences), 2003.

Battaglia, Andy. 'Queens Show Puts Trash in Full View—A Retrospective for the Department of Sanitation's Official Artist-in-Residence. *Wall Street Journal* (New York, NY; 23 September 2016): A.21.

Bauman, Zygmunt. 'Dream of Purity'. *Theoria: A Journal of Social and Political Theory*, No. 86. Dimensions of Democracy (October 1995): 49–60.

———. *Liquid Modernity*. Cambridge: Polity Press, 2000.

———. *Society Under Siege*. Cambridge: Polity, 2002.

———. *Wasted Lives: Modernity and Its Outcasts*. MA: Polity Press, 2012 [2004].

Beder, Sharon. 'Sydney's Toxic Green Olympics'. *Current Affairs Bulletin*, Vol. 70, No. 6 (November 1993): 12–18.

Benjamin, Walter. *The Writer of Modern Life: Essays on Charles Baudelaire.* Cambridge, MA: Harvard University Press, 2006.

Bijsterveld, Karin. *Mechanical Sound: Technology, Culture, and Public Problems of Noise in the Twentieth Century.* Cambridge, MA, London: MIT Press, 2008.

Biro, Andrew. *The Frankfurt School and Contemporary Environmental Crisis.* Toronto: University of Toronto, 2011: 43–72.

Bishop, Claire. *Artificial Hells: Participatory Art and the Politics of Spectatorship.* New York: Verso, 2012.

Blum, Elizabeth D. *Love Canal Revisited: Race, Class and Gender in Environmental Activism.* KS: University Press of Kansas, 2008.

Bourdieu, Pierre. *Logic of Practice.* Cambridge: Polity Press, 1992 (*Le sens pratique.* Paris: 1980).

Brechin, Steven R. Kempton, Willett. 'Global Environmentalism: A Challenge to the Postmaterialism Thesis?' *Social Science Quarterly*, Vol. 75, No. 2 (June 1994): 245–269.

Brenner, Neil. Keil, Roger (eds.). *The Global Cities Reader.* New York: Routledge, 2006.

———. 'Debating Planetary Urbanization: For an Engaged Pluralism'. Working Paper. Urban Theory Lab. Harvard GSD (Summer 2017): 1–23.

———. (ed.). *Critique of Urbanization: Selected Essays.* Basel: Birkhäuser, 2017.

Brookner, Jackie. 'The Heart of Matter'. *Art Journal*, Vol. 51, No. 2. On Art and Ecology (Summer 1992): 8–11.

Buchli, Victor. *Archaeologies of the Contemporary Past.* London and New York: Routledge, 2001.

Buchloh, Benjamin. 'Detritus and Decrepitude: The Sculpture of Thomas Hirschhorn'. *Oxford Art Journal*, Vol. 24, No. 2 (2001): 43–56.

Butler, Judith. *Precarious Life: The Powers of Mourning and Violence.* London: Verso, 2004.

———. *Frames of War: When Is Life Grievable?* London and New York: Verso, 2009.

———. 'Life, Vulnerability and the Ethics of Cohabitation'. *Journal of Speculative Philosophy*, Vol. 26, No. 2. Special Issue with the Society for Phenomenology and Existential Philosophy (2012): 134–151.

Campkin, Ben. 'Placing "Matter Out of Place": Purity and Danger as Evidence for Architecture and Urbanism'. *Architectural Theory Review*, Vol. 18, No. 1 (2004): 46–61.

———. *Remaking London: Decline and Regeneration in Urban Culture.* London: I.B. Tauris, 2013.

Carson, Rachel. *Silent Spring.* London and New York: Penguin Classics, 2000 [1962].

Casey, Edward S. *The Fate of Place: A Philosophical History.* CA and London: University of California Press, 1998.

Castoriadis, Cornelius. *The Imaginary Institution of Society.* Cambridge: Polity Press, 1987 [1975].

Cembalest, Robin. 'Talking Trash with Mierle Laderman Ukeles'. *The Jewish Daily Forward* (25 November 1994): 9.

Chakrabarty, Dipesh. 'Of Garbage, Modernity and the Citizen's Gaze'. *Economic and Political Weekly*, Vol. 27, No. 10/11 (7–14 March 1992): 541–547.

Chan, Jeffrey Kok Hui. 'The Ethics of Working with Wicked Urban Waste Problems: The Case of Singapore's Semakau Landfill'. *Landscape and Urban Planning*, Vol. 154 (October 2016): 123–131.

Chatman, Seymour. 'Voice'. *Narrative/Theory*. David H. Richter (ed.). New York: Longman, 1996.

Chi, Seung Won. 'Self-Identity and the Principles of Justice of Small Groups in a Community: In the Case of Nanjido Landfill Residents'. *Korean Social Theories*, Vol. 40 (Fall/Winter 2011): 151–180.

Choi, Jang Jip. *Democracy After Democratisation: Conservative Origin of Korean Democracy and Its Crisis*. Seoul: Humanitas, 2002.

Chung, Chae-sung. 'The Poverty of Nanjido Residents as Its Social Relations'. *Korean Society for Cultural Anthropology*, Vol. 21 (1989): 367–399.

Chung, Yeon-hee. *Nanjido*. Seoul: Jungeum Publishing, 1990.

Code, Lorraine. *Ecological Thinking: The Politics of Epistemic Location*. Oxford: Oxford University Press, 2006.

Conley, Verena A. *Ecopolitics: The Environment in Poststructuralist Thought*. London: Routledge, 1997.

Conte, Kari (ed.). *Mierle Laderman Ukeles: Seven Work Ballets*. Amsterdam: Kunstverein Publishing, 2015.

Cooke, Lynne et al. (eds.). *Robert Smithson: Spiral Jetty: True Fictions, False Realities*. Dia Art Foundation. Berkeley and London: University of California Press, 2005.

Corbin, Alain. *The Foul and the Fragrant: Odour and the French Social Imagination*. London: Picador, 1994.

Corner, James. *The Landscape Imagination: Collected Essays of James Corner, 1990–2010*. Alison Bick Hirsch (ed.). New York: Princeton Architectural Press, 2014.

Crysler, C. Greig. Cairns, Stephen. Heynen, Hilde (eds.). *The SAGE Handbook of Architectural Theory*. London: SAGE Publications, 2013 [2012]: 29, 501–512, 553–638.

Dauvergne, Peter. 'The Illusions of Environmentalism'. *The Shadows of Consumption: Consequences for the Global Environment*. Cambridge, MA: MIT Press, 2008: 209–218.

Davis, Frederick Rowe. *A History of Pesticides and the Science of Toxicology*. New Haven: Yale University Press, 2014.

Deleuze, Gilles. *Spinoza: Practical Philosophy*. trans. by Robert Hurley. CA: City Lights Books, 1988 [1970].

———. *What Is Philosophy?* trans. by Graham Burchill. London and New York: Verso, 1994 [1991].

Deleuze, Gilles. Guattari, Félix. *Anti-Oedipus: Capitalism and Schizophrenia*. trans. by Robert Hurley, Mark Seem and Helen R. Lane. London: Continuum, 2004 [1972].

Diken, Bülent. 'From Refugee Camps to Gated Communities: Biopolitics and the End of the City'. *Citizenship Studies*, Vol. 8, No. 1 (March 2004): 83–106.

Diken, Bülent. Laustsen, Carsten Bagge. 'Zones of indistinction: discipline, control and terror'. *Space and Culture*, Vol. 5, No. 3 (2002): 290–307.

Dillon, Brian. *Ruin Lust: Artists' Fascination with Ruins, From Turner to the Present Day*. London: Tate Publishing, 2014.

Douglas, Mary. *Purity and Danger: An Analysis of Concepts of Pollution and Taboo*. London and New York: Routledge, 2002 [1966].

Dudley, Nigel (ed.). *Guidelines for Applying Protected Area Management Categories*. Gland, Switzerland: IUCN, 2008, 2013: 1–24.

Dunlap, Riley E. Mertig, Angela G. 'Global Environmental Concern: An Anomaly for Postmaterialism'. *Social Science Quarterly*, Vol. 78, No. 1 (March 1997): 24–29.

Ek, Richard. 'Giorgio Agamben and the Spatialities of the Camp: An Introduction'. *Geografiska Annaler. Series B*, Vol. 88, No. 4 (2006): 363–386.

Emerson, R. W. 'First Visit to England' in *The Complete Works of Emerson Vol. V* (English Traits), Ch. 1; 'Plato; or, the Philosopher' in *The Complete Works of Emerson Vol. IV* (Representative Men), Ch. 2, and 'Art' in *The Complete Works of Emerson Vol. VII* (Society and Solitude), Ch. 3.

———. *Nature*. Boston: James Munroe and Company, 1836.

Fee, Elizabeth. Corey, Steven H. *Garbage! The History and Politics of Trash in New York City*. New York: New York Public Library, 1994.

Fela, Jen. 'Developing Countries Face E-waste Crisis'. *Frontiers in Ecology and the Environment*, Vol. 8, No. 3 (April 2010): 117.

Foster, Hal. *The Art-Architecture Complex*. New York: Verso, 2011.

Foucault, Michel. 'Questions on Geography'. *Power/Knowledge: Selected Interviews and Other Writings 1972–1977*. Colin Gordon (ed.). New York: Pantheon, 1980: 63–77.

———. 'Of Other Spaces'. trans. by J. Miskowiec, *Diacritics*, Vol. 16, No.1 (Spring 1986): 22–27.

———. *The History of Sexuality: The Will to Knowledge*. trans. by Robert Hurley. London: Penguin, 1998.

———. *The Birth of Biopolitics: Lectures at the Collège de France, 1978–1979*. trans. by Graham Burchell. Arnold I. Davidson (ed.). New York: Palgrave Macmillan, 2010 [2004].

Gandy, Matthew. 'Zones of Indistinction: Bio-Political Contestations in the Urban Area'. *Cultural Geographies*, Vol. 13 (2006): 497–516.

———. 'Landscapes of Disaster: Water, Modernity, and Urban Fragmentation in Mumbai'. *Environment and Planning*, Vol. 40, No. 1 (2008): 108–130.

———. *Fabric of Space: Water, Modernity, and the Urban Imagination*. Cambridge, MA: MIT Press, 2014.

———. 'From Urban Ecology to Ecological Urbanism: An Ambiguous Trajectory'. *Area*, Vol. 47, No. 2 (2015): 150–154.

Genosko, Gary (ed.). *The Guattari Reader: Pierre-Félix Guattari*. Oxford and Cambridge, MA: Blackwell Publishing, 1996.

———. *Deleuze and Guattari: Critical Assessments of Leading Philosophers Volume III*. London and New York: Routledge, 2001.

Gissen, David. *Subnature: Architecture's Other Environments*. New York: Princeton Architectural Press, 2009.

Gregory, Derek. 'Vanishing Points: Law, Violence and Exception in the Global War Prison'. *Violent Geographies: Fear, Terror and Political Violence*. Derek Gregory and Allan Pred (eds.). NY: Routledge, 2006a.

———. 'Defiled Cities'. *Singapore Journal of Tropical Geography*, Vol. 24, No. 3 (2006b): 307–326.

Guattari, Félix. *The Three Ecologies*. trans. by Ian Pindar and Paul Sutton. London and New York: Continuum, 2000 [1989].

Hadid, Zaha. Schumacher, Patrik. *Total Fluidity: Studio Zaha Hadid, Projects 2000–2010*. Vienna: University of Applied Arts Vienna, 2011.

Hall, Peter. *The World Cities*. New York: McGraw-Hill, 1966.

———. *Cities of Tomorrow: An Intellectual History of Urban Planning and Design in the Twentieth Century*. New York: Basil Blackwell, 1988.

———. 'Planning for the Mega City: A New East Asia Urban Form?' *East West Perspectives on 21st Century Urban Development*. John Brotchie et al. (eds.). Vermont: Ashgate, 1998.

Haraway, Donna J. *Simians, Cyborgs, and Women: The Reinvention of Nature*. London: Free Association Books, 1991.

———. 'Otherworldly Conversations, Terran Topics, Local Terms'. *Science as Culture*, Vol. 3, No. 1 (1992).

Hardt, Michael. Negri, Antonio. *Multitude*. New York: Penguin, 2004.

Harrison, Roy M. Hester, R. E. *Waste Treatment and Disposal*. Cambridge: Royal Society of Chemistry, 1995.

Harry, Joe. Harry, Joseph. 'Causes of Contemporary Environmentalism'. *Humboldt Journal of Social Relations*, Vol. 2, No. 1. Social Behaviour and Natural Environments (Fall/Winter 1974): 3–7.

Harvey, David. *Justice, Nature and the Geography of Difference*. MA: Blackwell Publishing, 1999 [1996].

———. *Rebel Cities: From the Right to the City to the Urban Revolution*. New York: Verso, 2013 [2012].

Hu, Wei et al. 'Emission Characteristics and Air-Surface Exchange of Gaseous Mercury at the Largest Active Landfill in Asia'. *Atmospheric Environment*, Vol. 79 (November 2013): 188–197.

Huh, Eun. 'Reconstruction Citizens Movement During the May 16 Military Regime: Integration and Division of the Citizens Movement in the Divided Nation'. *Historical Research*, No. 11. Special Issue on the Historicity of Park Chung-hee Era (December 2003): 11–51.

Hurlbut, Herbert S. Altman, Robert M. Nibley, Carlyle. 'DDT Resistance in Korean Body Lice'. *Science*, New Series, Vol. 115, No. 2975 (4 January 1952): 11–12.

Ingersoll, Richard. 'The Ecology Question'. *Journal of Architectural Education*, Vol. 45, No. 2 (2013): 125–127.

Inglehart, Ronald. *Culture Shift in Advanced Industrial Society*. NJ: Princeton University Press, 1990.

———. *The Silent Revolution: Changing Values and Political Styles Among Western Publics*. NJ: Princeton University Press, 2015 [1977].

Jameson, Fredric. 'Notes on Globalization as a Philosophical Issue'. *Cultures of Globalization*. Fredric Jameson and Masao Miyosi (eds.). NC: Duke University Press, 1998.

———. 'Globalization and Political Strategy'. *New Left Review* 4 (July–August 2000).

Ji, Joo Hyoung. 'Learning from Crisis: Political Economy, Spatio-Temporality, and Crisis Management in South Korea, 1961–2002'. Dissertation at Lancaster University, 2005.

———. *The Origin and the Creation of Neoliberalism in Korea*. Seoul: Chaek Sesang, 2011.

Ji, Soo-gul. 'Japanese Militarism, Fascism and Joseon [Korea] Farming Village Development Movement'. *History Criticism*, Vol. 47 (Summer 1999).

Joseph, Kurian. Visvanathan, C. 'Dumpsite Rehabilitation'. *Landfill Research Focus*. Ernest C. Lehmann (ed.). New York: Nova Science Publishers, 2007.

Kamaruddin, M. A. et al. 'An Overview of Municipal Solid Waste Management and Landfill Leachate Treatment: Malaysia and Asian Perspectives'. *Environmental Science and Pollution Research*, Vol. 24, No. 5 (25 October 2017): 1–33.

Kang, Hong-Bin. 'Urban Environment and Citizens: Toward a Human-Centred Urban Plan'. *City Issue*, Vol. 184 (December 1981): 59–75.

———. 'Slum Redevelopment and the Housing Stabilization of the City Poor'. *Gamsa*, Vol. 17 (March 1989): 40–45.

———. 'Planned Development of the "Creative Milieu" and Its Sustainability: Experiences from Seoul Digital Media City'. *Seoul City Research*, Vol. 11, No. 2 (June 2010): 259–267.

Kekes, John. 'Purity and Judgment in Morality'. *Philosophy*, Vol. 63 (1988): 453–469.

Kellner, Douglas. 'Theorising Globalization'. *Sociological Theory*, Vol. 20, No. 3 (November 2002): 285–305.

Kidd, Quentin. Lee, Aie-Rie. 'Postmaterialist Values and the Environment: A Critique and Reappraisal'. *Social Science Quarterly*, Vol. 78, No. 1 (March 1997).

Kim, Cheol-woon. 'Modern Transfiguration of Moral Training: The Individual Confined by the State'. *A Collection of Philosophy Treatises*, Vol. 2, No. 48. Saehan Philosophy Association (2007): 137–162.

Kim, Hyun-ji. 'Perception Gap Between Social Strata of Leisure Activity Diversification: Focusing on the Han River Park in Seoul'. Master's thesis at University of Seoul, 2013.

Kim, Seung-Hee. 'The Bio-Political Power and Its Limitations Through a Cholera Occurred in 1969 in South Korea'. *Social Thoughts and Culture*, Vol. 18, No. 1. Association of East Asian Social Thoughts (March 2015): 217–248.

Kim, Soon-jeon et al. *The Moral Training of the Imperial Japan: Research on 'The Textbook on Moral Training' Published by the Japanese General Government*. Seoul: J&C, 2008.

Kim, Yeok-ki. Jang, Se-hoon. 'Internal Differentiation and Reproduction of Poverty: Focusing on Nanjido Area'. *Korean Sociological Association*, Vol. 21 (April 1988).

King, Anthony D. *Global Cities*. London: Routledge, 1990.

Klein, Caroline et al. (eds.). *Regenerative Infrastructures: Freshkills Park, NYC: Land Art Generator Initiative*. Munich, London and New York: Prestel, 2013.

Klima, Ivan. *Love and Garbage*. trans. by Ewald Osers. New York: Vintage, 2002: 15–16.

Koolhaas, Rem. 'Evil Can Also Be Beautiful'. *Spiegel International* (27 March 2006). Accessed on 22 October 2013.

———. 'The Invention and Reinvention of the City'. *World Policy* (18 April 2012). Accessed on 22 October 2013.

Koolhaas, Rem. Cleijne, Edgar. *Lagos: How It Works*. Harvard Project on the City. Baden: Lars Muller Publishers, 2007.

Kracauer, Siegfried. 'Photography'. *The Mass Ornament: Weimar Essays*. Thomas Y. Levin (trans. and ed.). Cambridge, MA and London: Harvard University Press, 1995: 46–63.

Kunz, Robert G. *Environmental Calculations: A Multimedia Approach*. New York: John Wiley & Sons, 2010: 543–557.

Lacan, Jacques. *Écrits*. London: W.W. Norton, 2003 [1966].

Laginestra, E. et al. 'Towards Sustainable Urban Ecology, the Olympic Site Message'. *Geographical Education*, Vol. 14 (2001): 27–30.

Lawson, Benjamin Gordon. 'Garbage Mountains: The Use, Redevelopment, and Artistic Representation of New York City's Fresh Kills, Greater Toronto's Keele Valley, and Tel Aviv's Hiriya landfills'. PhD dissertation in History, University of Iowa, 2015.

Lee, Ho. 'The Residence Right of the Nanjido Residents'. *City and Poverty*, Vol. 21 (March 1996): 47–67.

Lee, Seung-won. Oh, Sun-bin. Chung, Yeo-wool. *Political Imagination of the Citizen-State*. Seoul: Somyung Publishing, 2003.

Lee, Young June. 'Art, Not for Creating but for Measuring' (On Joon Kim's artwork *Instant Landscape (Nanjido)*, 2012–2013). An essay for the SeMA Nanji Residency Exhibition 2013.

Lefebvre, Henri. 'The Right to the City' (Le droit a la ville), 1968.

———. *The Urban Revolution*. trans. by Robert Bononno. Minneapolis and London: University of Minnesota Press, 2003 [1970].

Lévi-Strauss, Claude. *The Savage Mind*. IL: University of Chicago Press, 1966 [1962].

Lim, Taehun. 'A Discovery of Soundscape' (On Joon Kim's 'Tempest' Project, 2015). An essay for the exhibition 'Time Collector' at the Gyeonggi Museum of Modern Art, 2015.

Lintott, Sheila. 'Ethically Evaluating Land Art: Is It Worth It?' *Ethics, Place & Environment*, Vol. 10, No. 3 (October 2007): 263–277.

Low, Setha. Taplin, Dana. Scheld, Suzanne. *Rethinking Urban Parks: Public Space and Cultural Diversity*. Austin, TX: University of Texas Press, 2005.

Lubick, Naomi. 'Shifting Mountains of Electronic Waste'. *Environmental Health Perspectives*, Vol. 120, No. 4 (April 2012): A148–A149.

Lugones, Maria. 'Purity, Impurity, and Separation'. *Signs* 19 (1994): 463.

MacCormack, Carol P. Strathern, Marilyn (eds.). *Nature, Culture and Gender*. Cambridge: Cambridge University Press, 1980: Chapter 1.

Madurapperuma, B. D. Kuruppuarachchi, K. A. J. M. 'Rehabilitating a Landfill Site of Lowland Tropical Landscape into an Urban Green Space: A Case Study from the Open University of Sri Lanka'. *International Journal of Sustainable Built Environment*, Vol. 5, No. 2 (December 2016): 400–410.

Marcuse, Peter. , VanKempen, Richard (eds.). *Globalizing Cities: A New Spatial Order?* Oxford: Blackwell, 2000.

Margaret J, Mary. 'Urban Parks by Hargreaves Associates'. *Envisioning the Millennium Park*. International Symposium Towards the Sustainable Development of Nanjido. Seoul, 1999.

Marshall, Douglas A. 'The Dangers of Purity: On the Incompatibility of "Pure Sociology" and Science'. *The Sociological Quarterly*, Vol. 49, No. 2 (Spring 2008): 209–235.

Massey, Doreen. 'A Global Sense of Place'. *Reading Human Geography: The Poetics and Politics of Inquiry*. T. Barnes and D. Gregory (ed.). London: Arnold, 1997: 315–323.

———. *World City*. Cambridge: Polity Press, 2007.

Matilsky, Barbara C. *Fragile Ecologies: Contemporary Artists' Interpretations and Solutions*. New York: Rizzoli International, 1992.

Mbembe, Achille. 'Necropolitics'. *Public Culture*, Vol. 15, No. 1 (2003): 11–40.

Mcevilley, Thomas. 'Philosophy in the Land'. *Art in America*, Vol. 92, No. 10 (November 2004): 158–163, 193.

Melucci, Alberto. *The Playing Self: Person and Meaning in the Planetary Society*. Cambridge: Cambridge University Press, 1996.

Merrifield, Andy. *Henri Lefebvre: A Critical Introduction*. London: Routledge, 2006.

———. *The Politics of the Encounter*. GA: University of Georgia Press, 2013 [2012].

———. *The New Urban Questions*. London: Pluto Press, 2014.

Meyer, Esther da Costa. 'Architectural History in the Anthropocene: Towards Methodology'. *The Journal of Architecture*, Vol. 21, No. 8 (2016): 1203–1225.

Mitchell, Don. 'The Annihilation of Space by Law'. *Antipode*, Vol. 29, No. 3 (1997): 303–335.

Moss, Jo. 'Redeveloping Homebush Bay'. *Envisioning the Millennium Park*. International Symposium Towards the Sustainable Development of Nanjido. Seoul, 1999.

Mostafavi, Mohsen. Doherty, Gareth (eds.). *Ecological Urbanism*. Basel: Lars Müller Publishers, 2010.

Mostafavi, Mohsen. Najle, Ciro (eds.). *Landscape Urbanism: A Manual for the Machinic Landscape*. London: Architectural Association, 2003.

Mumford, Mark. 'Form Follows Nature: The Origins of American Organic Architecture'. *Journal of Architectural Education*, Vol. 42, No. 3 (Spring 1989).

O'Neill, Alexander. 'What Globalization Means for Ecotourism: Managing Globalization's Impacts on Ecotourism in Developing Countries'. *Indiana Journal of Global Legal Studies*, Vol. 9, No. 2 (Spring 2002): 501–528.

Park, Jin-do. Han, Do-hyeon. 'New Town Movement and Yushin Regime (4th Republic): On Park Chung-hee Regime's Farming Area New Town Movement'. *History Criticism* (Summer 1999).

Park, Seo-young. 'Historical Transformations of Sangam-dong Based on the Lived Experience of Its Local Residents'. Master's thesis in Cultural Studies at Yeonsei University, 2004.

Park, Sun-joo. 'Nanjido Teleport Development Plan: Focusing on a Business Park Plan'. Master's thesis at Seoul National University (Graduate School of Environmental Studies), 1993.

Penny, Joe. 'Insecure Spaces, Precarious Geographies: Biopolitics, Security and the Production of Space in Jerusalem and Beyond'. DPU Working Paper, No. 141. UCL Development Planning Unit (2010).

Perkins, John H. 'Reshaping Technology in Wartime: The Effect of Military Goals on Entomological Research and Insect-Control Practices'. *Technology and Culture*, Vol. 19, No. 2 (April 1978): 169–186.

Plumwood, Val. *Feminism and the Mastery of Nature*. London and New York: Routledge, 1993.

Powell, Hilary. Marrero-Guillamón, Isaac (eds.). *The Art of Dissent: Adventures in London's Olympic State*. London: Marshgate Press, 2012.

———. *Urban Alchemy*, 2015.

Pugh, Michael. Rushbrook, Philip. *Solid Waste Landfills in Middle- and Lower-Income Countries: A Technical Guide to Planning, Design, and Operation*. Herndon, VA: World Bank Publications, 1999.

Purcell, Mark. 'Excavating Lefebvre: The Right to the City and Its Urban Politics of the Inhabitant'. *GeoJournal*, Vol. 58, No. 2. Social Transformation, Citizenship and the City (2002): 99–108.

Rahe, Paul A. *Montesquieu and the Logic of Liberty: War, Religion, Commerce, Climate, Terrain, Technology, Uneasiness of Mind, the Spirit of Political Vigilance, and the Foundations of the Modern Republic*. New Haven: Yale University Press, 2009: 99–102.

Rathje, William. Murphy, Cullen. *Rubbish! The Archaeology of Garbage: What Our Garbage Tells Us About Ourselves*. New York: Harper Collins, 1992.

Rawes, Peg (ed.). *Relational Architectural Ecologies: Architecture, Nature and Subjectivity*. London: Routledge, 2013.

Ray, Sarah Jaquette. 'Toward an Inclusive Environmentalism'. *The Ecological Other: Environmental Exclusion in American Culture*. AZ: University of Arizona Press, 2013: 179–184.

Reid, Donald G. 'Globalization and the Political Economy of Tourism Development'. *Tourism, Globalization and Development: Responsible Tourism Planning*. London: Pluto Press, 2003.

Relph, Edward. *Place and Placelessness*. London: Pion Books, 1976.

Reynolds, Ann. *Robert Smithson: Learning from New Jersey and Elsewhere*. Cambridge, MA and London: MIT Press, 2003.

Riley, Woodbridge. 'Transcendentalism and Pragmatism: A Comparative Study'. *The Journal of Philosophy, Psychology and Scientific Methods*, Vol. 6, No. 10 (May 1909): 263–266.

Rosenzweig, Roy. Blackmar, Elizabeth. *The Park and the People: The History of Central Park*. Ithaca, NY: Cornell University Press, 1998.

Rozman, Gilbert. 'The Northeast Asian Regional Context for Environmentalism: Assessing Environmental Goals Against Other Priorities in the 1990s'. *Journal of East Asian Studies*, Vol. 1, No. 2. Special Issue: Perspectives on Environmental Protection in Northeast Asia (August 2001): 13–30.

Russell, Edmund. *War and Nature: Fighting Humans and Insects with Chemicals from World War I to Silent Spring*. Cambridge: Cambridge University Press, 2001.

Sassen, Saskia. *The Global City: New York, London, Tokyo*. Princeton and Oxford: Princeton University Press, 2001 [1991].

———. *Expulsions: Brutality and Complexity in the Global Economy*. Cambridge, MA: Harvard University Press, 2014.

Scanlan, John. *On Garbage*. London: Reaktion Books, 2005.

Schama, Simon. *Landscape and Memory*. New York: A. A. Knopf, 1995.

Shin, Jieun. 'Waste and the Reconstruction of Urban Consumer Culture'. *Journal of Modern Social Science*, Vol. 17 (2013): 17–38.

Short, John R. *The Urban Order: An Introduction to Urban Geography*. Cambridge: Blackwell, 1995.

Soper, Kate. *What is Nature?: Culture, Politics and the Non-Human*. Oxford and Cambridge, MA: Blackwell Publishing, 1995.

Steiner, Frederick. 'Landscape Ecological Urbanism: Origins and Trajectories'. *Landscape and Urban Planning*, Vol. 100, No. 4 (2011): 333–337.

Strasser, Susan. *Waste and Want: A Social History of Trash*. New York: Henry Holt, 1999.

Sumner, J. *Co-operative Programme of Research on the Behaviour of Hazardous Wastes in Landfill Sites: Final Report of the Policy Review Committee*. London: H.M.S.O., 1978.

Tang, Yu-Ting. Nathanail, C. Paul. 'Sticks and Stones: The Impact of the Definitions of Brownfield in Policies on Socio-Economic Sustainability'. *Sustainability* (2012): 840–862.

Taylor, Liz. 'Place: An Exploration'. *Teaching Geography*, Vol. 30, No. 1 (Spring 2005): 14–17.

Tewdwr-Jones, Mark. Phelps, Nicholas A. Freestone, Robert (eds.). *The Planning Imagination: Peter Hall and the Study of Urban and Regional Planning*. London and New York: Routledge, 2014: 1–12, 217–227.

The Textbook on Ethics for Elementary School; Jong-hwa Ahn et al. *Modern Textbook on Moral Training I*. trans. by Jae-young Huh et al. (Seoul: Ewha Womans University Institute of Korean Culture, 2011).

The Textbook on Moral Training (*su-shin-gyo-gwa-seo*, 修身教科書) for Middle School (Seoul: Hwimungwan Publishing, September 1906).

The Textbook on Moral Training for Elementary School (Seoul: Dongmunsa Publishing, July 1909).

The Textbook on Moral Training for High School (Seoul: Hwimungwan Publishing, August 1907).

Thomas, Keith. *Man and the Natural World*. London: Penguin, 1991.

Thompson, Michael. *Rubbish Theory: The Creation and Destruction of Value*. Oxford: Oxford University Press, 1979.

Thompson, Paul. 'Building on Marginal and Derelict Land: Stockley Park London'. *Envisioning the Millennium Park*. International Symposium towards the Sustainable Development of Nanjido. Seoul, 1999.

Trepl, Ludwig. 'City and Ecology'. *Capitalism, Nature, Socialism*, Vol. 7 (1996): 85–94.

Tuan, Yi-Fu. *Space and Place: The Perspective of Experience*. MN: University of Minnesota Press, 1977.

Tufnell, Ben. *Land Art*. London: Tate, 2006.

Ukeles, Mierle Laderman. 'A Journey: Earth/City/Flow'. *Art Journal*, Vol. 51, No. 2. On Art and Ecology (Summer, 1992): 12–14.

US EPA. 'DDT Regulatory History: A Brief Survey'. EPA Report (July 1975). Accessed on 1 February 2016.

———. 'Decision-Making Guide to Solid Waste Management'. 1989.

———. 'Municipal Solid Waste Landfill Criteria, Subpart F—Closure and Post-Closure'. 1993.

Veolia Environment, S. A. *Report of Methane Treatment in Fudeken Sanitary Landfill*. Taiwan, R.O.C.: Environmental Protection Bureau, Kaohsiung City Government, 2014.

Vollgraff, Matthew. 'The Archive and the Labyrinth: On the Contemporary Bilderatlas'. *October*, Vol. 149 (Summer 2014): 143–158.

Wacquant, Loïc. *Urban Outcasts*. Cambridge: Polity Press, 2008 [2007].

Waldheim, Charles (ed.). *The Landscape Urbanism Reader*. New York: Princeton Architectural Press, 2006.

Wallerstein, Immanuel. *World-Systems Analysis: An Introduction*. NC: Duke University Press, 2004.

———. 'After Developmentalism and Globalization, What?' *Development Challenges for the 21st Century*. Cornell University (1 October 2004).

———. *The Modern World-System I: Capitalist Agriculture and the Origins of the European World-Economy in the Sixteenth Century*. CA: University of California Press, 2011 [1974].

Weller, Robert P. *Discovering Nature: Globalisation and Environmental Culture in China and Taiwan*. Cambridge: Cambridge University Press, 2006.

Weng, Yu-Chi et al. 'Proposal of an Integrated Evaluation Approach on Final Disposal Sites with Regard to Future Reclamation: A Case Study of Moerenuma Park, Sapporo City'. *Journal of Japan Society of Civil Engineers, Series G. II* (2013): 313–320.

———. 'Management of Landfill Reclamation with Regard to Biodiversity Preservation, Global Warming Mitigation and Landfill Mining: Experiences from the Asia–Pacific Region'. *Journal of Cleaner Production*, Vol. 104 (1 October 2015): 364–373.

Westendorff, David G. 'Three Essays on Urban Governance and Habitat in Developing Countries'. Dissertation at Cornell University, 2009.

Williams, Paul T. *Waste Treatment and Disposal*. Chichester: Wiley, 2005 [1998].

Wilson, Robin. 'At the Limits of Genre: Architectural Photography and Utopic Criticism'. *Journal of Architecture*, Vol. 10, No. 3 (2005): 265–273.

Wray, Timothy. Higgott, Andrew (eds.). *Camera Constructs: Photography, Architecture and Modern City*. London: Ashgate, 2012.

Yoo, Jae-soon. *The Nanjido People*. Seoul: Geulsure, 1989.

Yoon, Soo-jong. 'The Characteristics and Changes of Ragpickers Community'. *Democracy and Human Rights*, Vol. 2, No. 1 (2003a): 175–212.

———. 'Ragpickers and Nation: A history of Concentration of Ragpickers of South Korea'. *The Radical Review*, No. 56 (Summer 2003b): 265–296.

Zhu, Wei et al. 'Emission Characteristics and Air-Surface Exchange of Gaseous Mercury at the Largest Active Landfill in Asia'. *Atmospheric Environment*, Vol. 79 (November 2013): 188–197.

Zimmer, Anna. 'Urban Political Ecology: Theoretical Concepts, Challenges and Suggested Future Directions'. *Erdkunde*, Vol. 64 (2010): 323–354.

Interviews

Ahn, Gye-dong. CEO of Dongshimwon Landscape Design Company, Former member of the World Cup Park landscape design project (2001–2002) (11 August 2014).

(Anonymous). Senior Curator, National Museum of OOO (12 April 2014).

Chang, Kyung-Whan. Reverend, Lived in Nanjido Landfill, Residents' Representative (1980–2001) (13 August 2014).

Cho, Hyun-A. SeMA Nanji Art Residency Artist (17 April 2014).

Jung, In-hwa. Manager, Cleaning Department, Gangdong-gu, Seoul City, Former manager of vehicles in Nanjido Landfill (mid-1980s–early-1990s) (14 May 2014).

Kang, Deok-hee. Manager, Korea Federation for Environmental Movement (28 April 2014).

Kim, Joon. SeMA Nanji Art Residency Artist (3 August 2014).

Kim, Woon-soo (Dr). Senior Researcher, The Seoul Institute, Former Leading Researcher of *Evaluation of Nanjido Landfill and Environmentally-friendly Restoring Strategies* 2000 (22 April 2014).

Kim, Yong-soo. Manager, Water Treatment Division, Seoul City, Former Manager of Nanjido residents' Moving (early 2000s) (16 May 2014).

Lee, In-sung. Professor of Department of Landscape, University of Seoul, Former Director of the World Cup Park Landscape and Development Project (2001–2002) (27 August 2014).

Lee, Jong-tae. Chair, Sangam-dong Community (7 May 2014).

Lee, Min-jung. Assistant, Korea Federation for Environmental Movement (28 April 2014).

Lee, Wonho. SeMA Nanji Art Residency Artist (2 March 2017).

Lim, Jade Keunhye. Head of the Exhibition Team, National Museum of Modern and Contemporary Art, Seoul (Former Senior Curator, Seoul Museum of Art) (12 April 2014).

Magdalena. Nun, Lived in Nanjido Landfill for Four Years (late 1980s) (26 August 2014).

Park, Sae-bom. Manager, West Seoul Parks and Landscape Management Office, Seoul City (14 May 2014).

Park, Woong-bin. Director, West Seoul Parks and Landscape Management Office, Seoul City (14 May 2014).

Vantu, Quynh. SeMA Nanji Art Residency Artist (7 March 2017).

Yoo, Jae-soon. Novelist & Journalist, Lived in Nanjido Landfill for Several Years (early 1980s) (21 August 2014).

Magazine essays

Choi, Byung-cheon. 'Nanjido Report: Nanjido People's Lives for Survival'. *New Family*, Vol. 369 (May 1987): 34–42.

Kang, Hyun. 'Waiting for Love'. *New Family*, Vol. 386 (December 1988): 100–103.

Kim, Doo-hee. 'Nanjido Development Possible Only After 15 Years'. *Science Dong-A*, Vol. 518 (August 1990): 126–129.

Lee, Cheol-Jae. 'For Whom the Nanjido Golf Course Is Built?' *Ways to Live Together*, Vol. 96. Korean Federation for Environmental Movements (June 2001).

Lee, Chul-Jae. 'Nanjido, Would It Be the Korean Love Canal?' *Ways to Live Together*, Vol. 100. Korean Federation for Environmental Movements (October 2001).

Lee, Ho. 'The Residence Right of the Nanjido Residents'. *Urbanity & Poverty*, Vol. 21 (March 1996): 47–67.

Lee, Sun-hee. 'Seoul Landfill, Nanjido: The Falling Stars'. *Saemteo*, Vol. 19, No. 12 (December 1988): 10–14.

LeeYou, Ju-hye. 'Painting World in Nanjido Landfill'. *Saemteo*, Vol. 32, No. 1 (January 2001): 59.

Nang, Hyo-sik. 'One Afternoon of Nanjido'. *New Family*, Vol. 381 (June 1988): 114–115.

Shin, Chan-Gyun. 'Nanjido, The Sound of Hoeing'. *Gonggan*, Vol. 124 (October 1977).

Shin, Il-Cheon. 'Democratic Republic Established by the Juvenile Vagabonds'. *Sasanggye*, Vol. 10, No. 2 (1962).

Song, Un. 'The Last Winter of Nanjido Children'. *Mid-Level Education*, Vol. 24 (February 1992): 112–117.

Newspapers

Korean News on the Nanjido Region from 1953 until 2015

Korean News on DDT and Sanitation Policies from 1948 until 1988

'China Slows Garbage Imports and Impact Is Felt Globally' (Video). *The New York Times* (25 November 2017).

'China's Trash Ban Forces Europe to Confront Its Waste Problem' (from Getting Wasted series). *Politico EU* (21 February 2018).

Freytas-Tamura, Kimiko, de. 'Plastics Pile Up as China Refuses to Take the West's Recycling'. *The New York Times* (11 January 2018).

Gordon, Sarah. 'Shocking Images Show Horror of Child Labor at Toxic Cambodia Rubbish Site'. www.dailymail.co.uk (16 March 2015). Accessed on 1 May 2018.

Hilaire, Eric. 'Photographer Zooms in Beijing's Waste'. *The Guardian* (26 March 2010).

Kim, Sung-jong. 'A Secret Paradise'—Photographs of Dong-hoon Sung at Gallery ryugaheon, (Seoul, 2015). www.photomarketing.co.kr (19 November 2015). Accessed on 1 May 2018.

Taylor, Matthew. 'Rubbish Already Building up at UK Recycling Plants Due to China Import Ban'. *The Guardian* (2 January 2018).

Seoul City government's documents

The Ministry of Construction, South Korea. 1967

Seoul City. *A Preliminary Plan for the Construction of Levee Between Nanjido and Haengju Fortress*. 1980.

———. (Headquarter of Cleaning Projects). *A Preliminary Plan for Overground Landfill of Nanjido*. 1985a.

———. (Department of Cleaning, Environmental Administration). *A Report on the Environmental Impact of the Overground Landfill of Nanjido*. 1985b.

———. (Headquarter of Construction). *A Report on the Site-Visit of Foreign Waste Treatment Facilities—Related to Nanjido Landfill*. 1985c.

———. *A Preliminary Urban Planning of Seoul City*. 1990.

———. *A Preliminary Plan for Environmental Protection and Nanjido Landfill's Stabilisation*. 1992a.

———. (Daewoo Engineering Co, Ltd.). *A Preliminary Report on the Long-Term Land Use Plan for Nanjido Landfill*. 1992b.

———. (Samsung Construction Co. Ltd.). *A Study on the High-density Business Town Development of Seoul*. 1992c.

———. (Headquarter of Environmental Management). *A Plan for the Nanjido Landfill Stabilisation Construction*. 1996.

———. 'The Processes of the Relocation of the Nanjido Collective Housing Complex Residents' (document). August 2001.

———. (Headquarter of Construction and Safety). *From Garbage Mountain to the Land of Life*: Proceedings of Nanjido Landfill Stabilisation Construction. 2003a.

———. (Park and Management Office) (Sunjin E&A). *Making of the World Cup Park*. 2003b.

Seoul Foundation for Arts and Culture (ed.). 1995 Seoul, Sampoong: *A Record of the Sampoong Department Disaster for Social Memory*. 2016.

The Seoul Institute. *Evaluation of Nanjido Landfill and Environmentally-Friendly Restoring*. 2000.

The Seoul Institute and University of Seoul. *Envisioning Millennium Park*. Proceedings of the International Symposium 'Towards the Sustainable Development of Nanjido' (1–3 December). 1999.

Films and videos

'Nanjido'. MBC TV documentary. 1993.
'Nanjido, Embracing Life'. EBS TV documentary. 2009.
'Reviving Nanjido'. KBS TV documentary. 1994.
Varda, Agnès. *The Gleaners and I (Les glaneurs et la glaneuse)*. 2000.
Walker, Lucy. Harley, Karen. *Waste Land*. 2010.

Index

Page numbers in **bold** reference tables.
Page numbers in *italics* reference figures.